Understanding Science 2

Joe Boyd
Head of Chemistry, Beeslack High School, Penicuik
and
Walter Whitelaw
Assistant Science Adviser, Lothian

JOHN MURRAY

Advisory panel
Peter Leckstein: *Inspector for Living Science, London Borough of Wandsworth*
Lesley Campbell: *Health Education Adviser*
Terry Allsop: *Lecturer in Education Studies, Oxford University*

First published 1990
by John Murray (Publishers) Ltd
50 Albemarle Street, London W1X 4BD

Reprinted 1990, 1991 with revisions, 1992

Designed by Impress International, 33540 France
Typeset by Blackpool Typesetting Services Ltd, Blackpool
Printed in Great Britain by
Butler & Tanner Ltd, Frome and London

British Library Cataloguing in Publication Data
Boyd, Joe
  Understanding science book 2.
  Pupils' bk
  1. Science
  I. Title   II. Whitelaw, Walter
  500

ISBN 0-7195-4623-0 Pupils' book
ISBN 0-7195-4624-9 Teachers' resource book

Contents

continued ▶

Lively places

Describing places

A Around your school

Read this letter from a penfriend who lives in a surprising place.

Dear Friend,
Would you like to be my penfriend? I live in a wooden home which is 4 m long. It has a low ceiling, about 5cm high, and is very dark inside. The outside wall is painted light green but there is no paint on the inside walls and no carpet on the floor.
The temperature is a comfortable 20°c all year round. Wind, sleet and snow are no problem and the rainfall is 0mm per year! It is 100m above sea level. The main entrance faces north and the living room runs from east to west.
I live with my family. There are 14 of us. We live with three spiders, a few big beetles and a woodlouse who is awaiting a letter from the housing department. There are also some dangerous animals that roam around just outside our home.
Our house has one unusual feature. Every so often a steel machine appears at the front entrance with a bit of cheddar cheese balanced on it. No-one in the family knows what it is or where it comes from. We just leave it alone. None of us likes cheese anyway.
                                        Your friend,
                                         Jerry.

The letter tells you about
  ● the **appearance** of the place
  ● **measurements** describing the place
  ● the **living things** in the place
  ● the **effect of humans** on the place.
Can you guess where this penfriend lives?

---

1 Your group will study a **small** part of the school grounds. Imagine that this place is your home. Your penfriend wants to know about it.
2 Go to the place. Have a good look around. Make some observations about it.
3 Make notes on
  ● appearance   ● measurements
  ● living things   ● effect of humans.

---

1 Write a letter to your penfriend to describe your imaginary home.
2 **Collect** coloured pencils and drawing paper.
Make a poster with words and pictures to describe this part of the school grounds.
Include a photograph of it if possible.

CHECKPOINT

B Around and about

The different places in the school grounds are examples of different environments. An **environment** means a place, including the living things in it.

Here are some different environments.

Town street and gardens

Parkland

Rocky seashore

Forest

Arable farmland

River bank

Choose the three environments that you know most about.
Write about each environment.
Describe the place and the living things in it.

CHECKPOINT

Measuring the environment

A Place measurements

A poster or a photograph is a good way to show the appearance of an environment. Measurements improve the description.

Some useful measuring devices

Collect

Instruments

> 1 Discuss how to use the instruments available.
> 2 Prepare a table to record your measurements.
>   The headings should be
>   ● Instruments
>   ● What we measured
>   ● Our result.
> 3 Go to one of the places in the school suggested by your teacher.
>   Make and record your measurements.

1 Write about the place you studied.
   Include all the measurements you made.
2 Which of your measurements would change quite often?
3 Which of your measurements would be useful in a weather report?
4 Which of your measurements would be different in six months' time?

CHECKPOINT

B Map measurements

Some measurements have been made already by scientists and used to make maps.

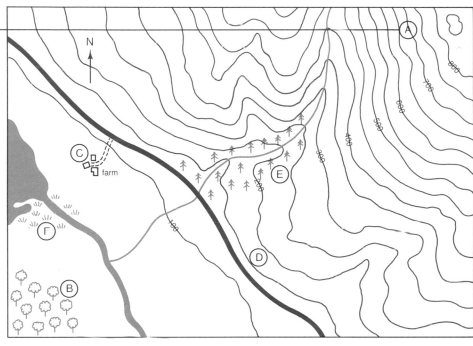

SITE A

Environment type: Hillside
(use clues from map)

Altitude: 700 m
(use contour lines)

Slope: Steep
(how close together are
contour lines?)

Direction of slope down: West
(look at contour lines)

Key

~ contours (50m apart)
━ road
↟ conifer trees
♧ broad-leaved trees
〜 stream
╲╿⁄ marsh

The types of plants that can grow in a place depend
- on the altitude and slope   - on the soil.

Site A	Site B	Site C	Site D

1 Describe the places marked E and F on the map above. Include measurements of altitude and direction of slope. Note the steepness of the slope.

2 If an Ordnance Survey map is available then describe your school's environment in the same way as you did for E and F.

CHECKPOINT

Looking at weather

A   Weather measurements

Weather affects the environment. The weather can change during a single day, and from day to day throughout the year. The weather map is for a day in winter. The key explains what the symbols mean.

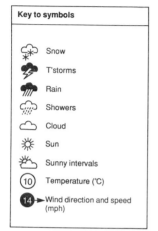

Key to symbols

☁❄	Snow
⛈	T'storms
🌧	Rain
🌧	Showers
☁	Cloud
☀	Sun
⛅	Sunny intervals
(10)	Temperature (°C)
(14)→	Wind direction and speed (mph)

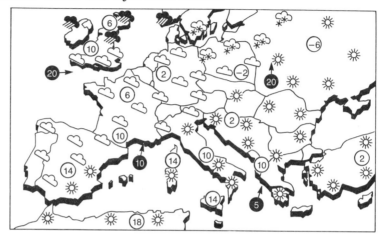

Collect

Weather chart
Wind chart
Coloured pencils
Measuring
instruments

1 Discuss with a partner how you could make the weather measurements shown on the map. Think carefully about how to measure wind direction.
2 Make as many measurements of today's weather as you can.
   Record your measurements on your weather chart. Use the same symbols as in the key.
3 Continue the record in your next four science lessons.

1 Look at the weather map above. What was the weather like on this day
   ● in the UK?
   ● in the Mediterranean region?
   Describe the weather in these places in words.
2 Copy and complete the table below to show how you would expect the weather to change in your area over a year. Describe the weather with symbols.

Date	Today	4 months from today	8 months from today
Temperature Wind Sky			

3 Animals and plants are affected by the weather. What possible reasons are there for the following observations?

Worms come to the surface on wet days

Some animals hibernate during winter

Trees don't grow on high ground

B Weather and farming

The pattern of weather for a region is called its **climate**. The climate affects the kind of animals that a farmer can keep. It also affects when seeds are planted and when it is time to harvest them.

Sometimes there are sudden unexpected changes in the weather that cause natural disasters.

Heavy rain

High winds

Drought

1 Why is climate important to farmers?

2 Describe how two types of disaster caused by bad weather affect a farmer.

3 Make a poster about how human activity is changing our climate. Use the books and magazines in the classroom or library. Key words to look for in the index are *climate*, *ozone layer* and *greenhouse effect*.

Counting living things

A Sampling animals

An environment can be changed by weather, natural disasters, human activity and by other living things. If the change is too great then some kinds of animals and plants may become **extinct** (die out).

Dinosaurs became extinct long ago. Cause: environmental change?

The Great Auk became extinct in 1844. Cause: human activity

Ospreys are in danger of becoming extinct in Britain. Cause: human activity

Sometimes one type of animal or plant becomes too numerous and upsets the environment for the others.

Deer remove the bark from trees

Rhododendrons stop other plants from growing

The numbers of animals and plants in an environment can be estimated. You can use this information to decide which are in danger.

Collect

Tray of sawdust
Magnetic rod
Sampling square

There are iron 'fish' hidden in the sawdust 'pond'. You are going to use a sampling square (called a quadrat) to estimate the number in the pond.

1 Put the sampling square anywhere on the tray.
2 Use the magnet to catch **all** the fish **inside** the sampling square. Count the fish.
3 Do this **two** more times. Each time, sample in a **different** place.

1 Make a table of your results. The two columns should have the headings
  ● Sample
  ● Number of fish caught.
2 Did you catch the same number of fish each time? Explain this.
3 Work out your **average** number in a sample. (*Hint:* add up the numbers and divide by three.)
4 It would take ten sampling squares to cover the whole pond. Use your average number in a sample to estimate (work out) how many fish are in the whole pond.

B Sampling plants

Collect

Sampling square

1 Find a piece of grassy land.
2 Choose one type of small plant that seems to be common in the place. (A daisy, for example, but not grass.)
3 Use the sampling square to estimate the number of these plants. Take three samples.

1 What was the size of your sampling square?
2 Work out the average number of plants in a sample.
3 Estimate the number of plants in 100 m^2 of this type of land.

Feeding living things

A Food chains

There may be many living things in an environment. These living things are always linked together in **food chains.**

All food chains have a similar pattern.

1 Light energy from the sun is trapped by green plants.
The light energy is changed into stored energy. This change is called **photosynthesis**.

2 The plant uses this energy to grow and stay alive until it is eaten by an animal or it dies.

prey

3 The animal uses the energy from the plant. The animal may become the **prey** of another animal. When it is eaten, energy is passed to the **predator**.

predator

Collect

Food-chain game
1 die
Red and green
coloured pencils

Play the food-chain game with a partner. The rules are on the game board.

1 What is the starting point for all food chains?
2 Explain what the words *predator* and *prey* mean.
3 What is the main change that takes place in photosynthesis?
4 Which living things carry out photosynthesis?
5 Why is photosynthesis very important?

B A sunny start

You are part of many food chains. You eat food. Energy in food comes from a plant or an animal. This energy can always be traced back to the sun.

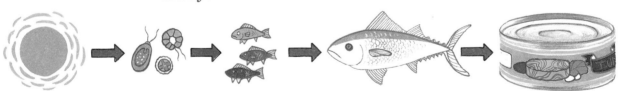

Write down the food chains for three of the foods on the buffet table.

Helping and harming the environment

A Conservation and pollution

People can affect the environment in good and bad ways. **Conservation** is helping the environment. **Pollution** is harming the environment. The photographs below show examples of conservation and pollution.

a

d

b

c

e

f

Discuss the pictures. Which show conservation? Which show pollution?

1 Pick two pictures showing conservation. Describe what is happening in each.
2 Pick two pictures showing pollution.
   a What is causing the pollution?
   b What effect is the pollution having?
   c How do you think the pollution could be prevented?
3 Write about a local conservation activity **or** a local pollution problem.
   (Your teacher will give you a local newspaper to help you.)

CHECKPOINT

B Pollution detective

You can be a pollution detective. There are clues in the environment which indicate the amount of air pollution. Some plants and animals are good indicators of pollution.

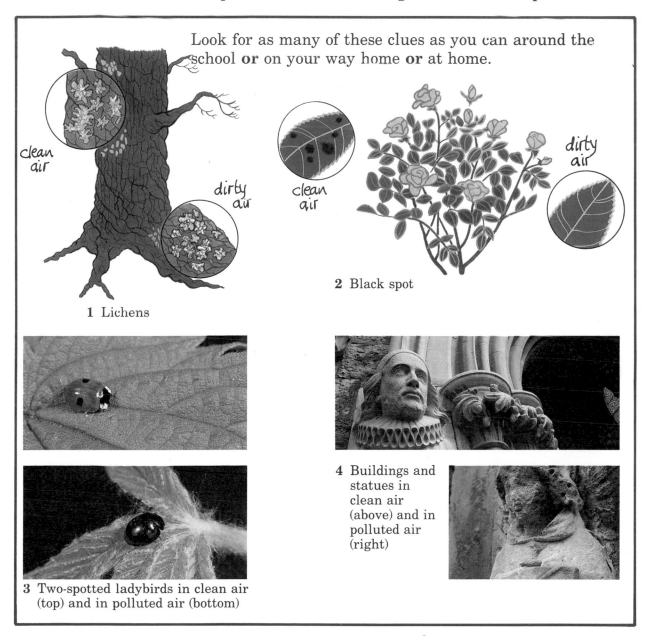

Look for as many of these clues as you can around the school **or** on your way home **or** at home.

clean air

dirty air

1 Lichens

clean air

dirty air

2 Black spot

3 Two-spotted ladybirds in clean air (top) and in polluted air (bottom)

4 Buildings and statues in clean air (above) and in polluted air (right)

1 Write a short report about your detective work.
2 Where do you think air pollution comes from? Make a list of sources.

CHECKPOINT

Problem

Winning an award

You can help to protect the balance of nature in your environment. Take part in the **Conservation Award Scheme**. Everyone's a winner:
- You will win a certificate
- The animals and plants in your area will win your care and attention.

1 Complete as many of the tasks as you can on your own or as part of a class group. You can take as long as you want.
   For each task
   - find out **how** to do it
   - make a **plan**
   - pick the best time and **act** on your plan.

2 Collect and complete an application form for an award. You need to earn
   - 600 points for the **gold** award
   - 350 points for the **silver** award
   - 100 points for the **bronze** award.

Fact-finding tasks

1 Find the addresses of four local conservation groups. **10 points**
2 Write to a local conservation group asking for information. **10 points**
3 Join a conservation group. **50 points**
4 Go round a nature trail. **50 points**
5 Watch a TV programme on nature **or** read a book about nature. **10 points** *per programme or book (up to 100 points)*
6 Collect articles from the local newspaper about local conservation issues. **10 points** *per article (up to 50 points)*

Fact-giving tasks

1 Set up a conservation display for the school library.
100 points
2 Organise a talk from a local conservation group at your school. **100 points**
3 Organise a visit to a local nature reserve.
150 points
4 Write an article for the school newspaper or for the local newspaper about nature conservation.
50 points *for school newspaper*
150 points *for local newspaper*

Action tasks

1 Collect litter from around the school. **50 points**
2 Take part in a clean up campaign in your area.
100 points
3 Make and use a bird table.
50 points *for each bird table*
(up to **200 points***)*
4 Make a wild garden. You can buy seeds to help you. **150 points**
5 Plant a tree with your classmates. Look after it.
100 points *per tree*
(up to **300 points***)*
6 Set up a tree nursery. (You can do this in a window box.) **300 points**
7 Make a nest for insects that live in holes. Use drinking straws or hollow plant stems, or make holes in dead wood. **50 points**
8 Plant a plot of small shrubs like hawthorn and holly. **50 points** *per type of shrub*
(up to **200 points***)*
9 Raise some money for a local or national conservation group. **5 points** *per pound*
10 Plan and carry out a survey of wild birds in your area. **10 points** *per type of bird spotted*
(up to **500 points***)*

Making an impact

The photograph shows a typical rural environment in Britain.

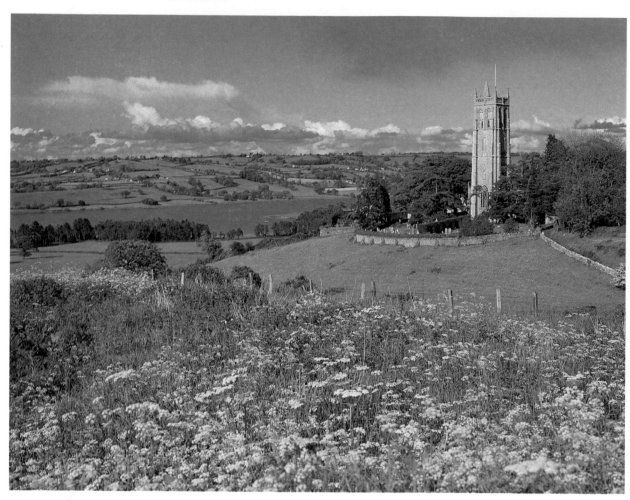

How do you think building each of the following would change this environment?

Nuclear power station

Leisure complex

Motorway

Readabout

Forming the environment

Landforms

The land around us has formed over millions of years. Movements of the Earth's crust, erupting volcanoes and earthquakes have shaped our surroundings.

The Earth's crust is made up of several giant slabs called plates that are moving very slowly. The Himalayas have been created over a long time as two of these plates meet and push against each other.

A volcano can erupt suddenly and make a new landform. The island of Surtsey, off Iceland, was produced in this way in 1963.

New landforms are changed slowly by weathering and erosion. Rocks are broken into smaller fragments by **weathering**. This can be caused by changes in temperature which cause water in cracks in the rock to freeze and thaw. It can also be due to chemical changes in the rock.

Water, wind or ice can then move the rock fragments. This is called **erosion**. The rounded pebbles on a beach or in a river bed show good evidence of movement and wearing away. Over a very long time, rugged mountains can be changed into rolling hills.

continued ▶

Soil

Soil is made up of bits of rock of different sizes as well as decaying plant and animal material. It holds water and air, and is the home for many living things. The bits of rock are formed from the weathering of the rock beneath. The type of soil which forms depends on the size of the grains in this rock and on the chemicals present in it.

For example, chalky soils are formed from the weathering of chalk. Grasses grow well on this type of soil. Many plants, heathers for example, do not grow well on it.

Soils used for growing crops can be improved by adding artificial fertilisers. The table shows some of the advantages and disadvantages of using these fertilisers.

Advantages	Disadvantages
● Crops can be grown on poor soil ● Crops grow more quickly ● A larger crop can be grown	● Fertilisers are expensive ● A lot of the fertiliser is washed away by rain. ● Fertilisers cause chemical pollution of streams and rivers ● Use of fertilisers can lead to land being over-used and the soil being ruined

1 How have the Earth's landforms been made?
2 a What causes weathering?
   b What causes erosion?
   c What effects do weathering and erosion have on the landscape?
3 Why are there different types of soil?
4 Imagine you are a farmer. Explain why you want to use artificial fertilisers.
5 Imagine you are concerned about the environment. Write a short letter to the farmer explaining the dangers of using artificial fertilisers.

2

Substances

A Oxygen

We describe things to other people every day. You might describe this lesson by saying that 'science was wonderful as usual'. This is your **opinion**. You might describe your teacher as 'the woman with the red jacket' or as 'the fellow with the rusty car'. These are **facts** (if they are true!). A really good description will be full of useful details.

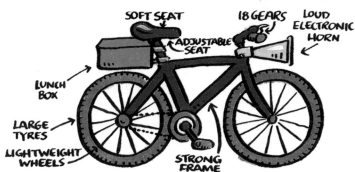

For example, an advert for a hi-tech bike might describe it with a diagram like this.

Some things are more difficult to describe. You are going to find out about an **invisible** gas called oxygen. You will describe something that you cannot even see.

Collect

2 stoppered tubes of oxygen
Bunsen burner
Heatproof mat
Wooden splints
Test tube holder
Safety glasses

Do experiments **a**, **b** and **c**.
Use the books available to answer **d** and **e**.

a What is its **appearance**?

b Does it **burn**?
Put a **lighted** splint into the oxygen.
What happens?

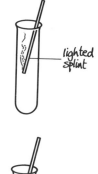

lighted splint

c Does it **help things to burn**?
Put a glowing splint into the oxygen.
What happens?

d What is its **melting point**?

e What are its **important uses**?

glowing splint

Write a good scientific description of oxygen.
Include everything that you have discovered.

CHECKPOINT

B Hydrogen

Hydrogen is another invisible gas. It has many uses.

Rocket fuel

Making margarine

Filling weather balloons

Find out about the appearance, burning, melting point and any other uses of hydrogen.

Write a good scientific description of hydrogen.
Include everything that you have found out about it.

Organising information

A Introducing metals

A long description is easier to understand if the information is organised in some way. We will put details about substances into three sets:

- **appearance**: what the substance looks like
- **properties**: what the substance does
- **uses**: what the substance is used for.

Metal X is common in homes. Perhaps you will recognise it from this description.

```
METAL X

Appearance:

Red/brown solid
Shiny and smooth
Sharp edges

Properties:

Melts at 1077 ºC
Boils at 2567 ºC
Good conductor of electricity and heat
Soft metal, can be cut by scissors
1cm³ weighs 8.9 g
Blackens when heated in air
Nothing happens when put in hydrochloric acid

Uses:

Electrical wiring, jewellery, coins, water pipes,
mixed with other metals to make alloys like
bronze and brass
```

Collect

2 strips of iron
2 strips of magnesium
Battery
Bulb
3 connecting wires

1 Do all the following investigations, first with iron and then with magnesium. Write down your results.

    a What is its appearance?

    b What is its mass?

    c What is its melting point? (You'll need to look this up.)

    d How easily does it break?

    e Is it a good conductor of heat?

    f Is it a good conductor of electricity?

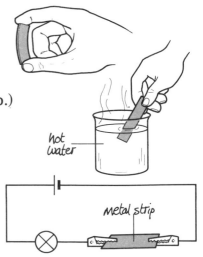

hot water

metal strip

Collect

Dilute hydrochloric acid
Test tube
Bunsen burner and mat
Tongs
Safety glasses

2 Use **one** strip of each metal for **each** of the following experiments.

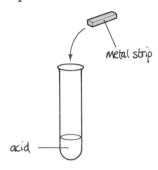

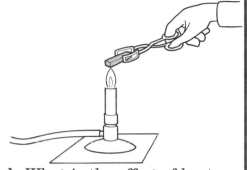

a What is the effect of acid on it?

b What is the effect of heat on it?
Heat the metal strongly for about a minute. **Do not look directly at the flame.**

1 Read the description of metal X again.
Write similar descriptions for iron and for magnesium.
Remember to put your results under the headings *Appearance* and *Properties*.

2 From your results, write down one possible use for each metal.

B An investigation

Collect

2 strips of copper
2 strips of zinc
Safety glasses
Anything else you need

Find out the properties of the metals zinc and copper.

Write a description for each of these metals.

Periodic table

A The table of elements

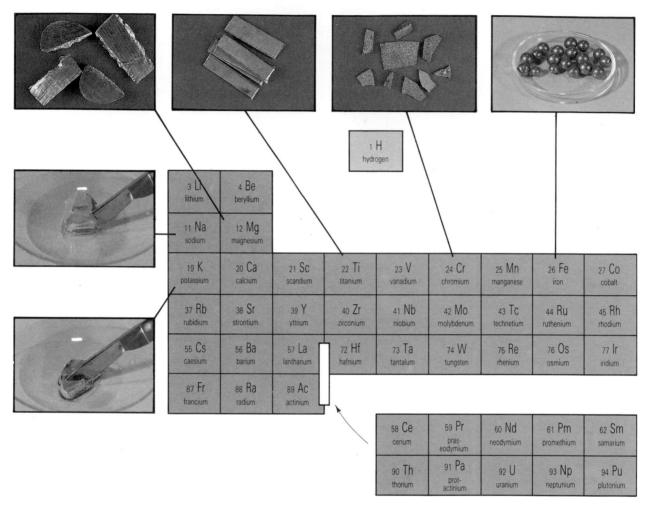

| 1 H hydrogen | | | | | | | | |

| 3 Li lithium | 4 Be beryllium |
| 11 Na sodium | 12 Mg magnesium |

19 K potassium	20 Ca calcium	21 Sc scandium	22 Ti titanium	23 V vanadium	24 Cr chromium	25 Mn manganese	26 Fe iron	27 Co cobalt
37 Rb rubidium	38 Sr strontium	39 Y yttrium	40 Zr zirconium	41 Nb niobium	42 Mo molybdenum	43 Tc technetium	44 Ru ruthenium	45 Rh rhodium
55 Cs caesium	56 Ba barium	57 La lanthanum	72 Hf hafnium	73 Ta tantalum	74 W tungsten	75 Re rhenium	76 Os osmium	77 Ir iridium
87 Fr francium	88 Ra radium	89 Ac actinium						

| 58 Ce cerium | 59 Pr praseodymium | 60 Nd neodymium | 61 Pm promethium | 62 Sm samarium |
| 90 Th thorium | 91 Pa protactinium | 92 U uranium | 93 Np neptunium | 94 Pu plutonium |

Scientists describe oxygen, iron and magnesium as elements. An element is made of only **one** substance. There are 92 natural elements and some others that have been made by people. All the elements are listed in a table called the **periodic table**.

1 In the periodic table, all the elements have a symbol. Why do you think these symbols are used?
2 Copy the following list, then underline the elements in it.
- water
- oxygen
- silver
- vinegar
- zinc
- salt
- bread
- air
- neon
- sugar
- brass
- carbon

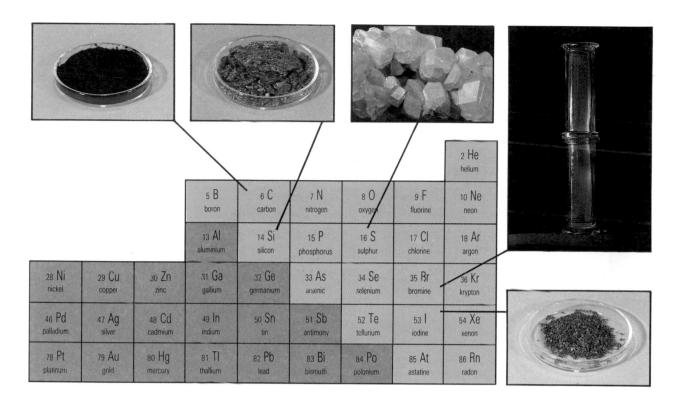

atomic number — symbol — 11 Na — sodium — name

3 List the names and symbols of five elements that you recognise from the periodic table.
Write down some facts about each one.
4 Where are the metals in the periodic table?

CHECKPOINT

B A chemical code

Use the table to crack this code. Write the message in English.

Scandium+Iodine+(Europium−Uranium)+Nitrogen+Cerium Iodine+Sulphur   Fluorine+Uranium+Nitrogen.

Now try to write your own short coded message.

CHECKPOINT

Elements make compounds

A Making compounds

Many useful things in our world are built from simpler parts.

For example, your home is built from materials like brick, cement and wood. You can only live in it because the materials have been joined together. These words are built from letters which have been joined together.

In a similar way, everything in the world is built from elements. The elements can join together to make new substances.

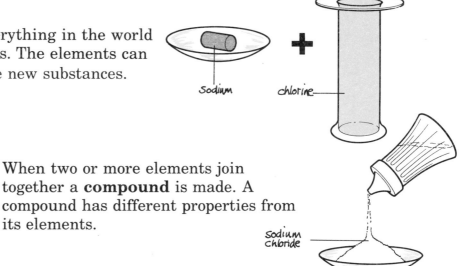

When two or more elements join together a **compound** is made. A compound has different properties from its elements.

Your teacher will make the compound copper chloride by joining the elements copper and chlorine together.

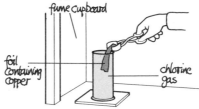

Now you can make a compound yourself.

1 Use one piece of water-indicator paper to find out what water does to it.
2 **Pop** the hydrogen with a lighted splint.
3 Test the tube with water-indicator paper. Look carefully at the edges of the paper.

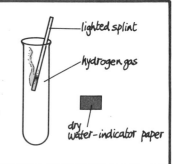

1 What is a compound?
2 Describe your experiment. Include a note of what you did and what happened.
3 Why did you begin with a **dry** test tube containing hydrogen?
4 Copy and complete the following summary (which is called a word equation).

h _____   and   oxygen   join to make   w _____

element   +   e ____ $\longrightarrow$ c _____

B Breaking compounds

A compound contains two or more elements joined together. If the join is broken then the elements may become free. The join can sometimes be broken by electricity.

Collect

Green solid
2 carbon rods
Beaker
Connecting leads
Crocodile clips
Power pack

Part 3:
Water
Bunsen burner and mat
Wooden splints
Gas-collecting equipment

1 Dissolve a spatula-ful of green solid in about 30 cm³ of water.
Set up the apparatus in the diagram.
Switch on.
2 Identify any elements by colour or smell (take great **care** to sniff very gently).
3 If you have time, repeat the experiment with water in place of the green solid.
Find a way to collect the gases which are produced.
Test them with a **lighted** splint.

1 Describe your experiment(s).
Include a drawing of the apparatus, and a note of what you did and what happened.
2 What do you think the elements in the green solid are?
3 Write a summary (word equation) for your experiment(s).

2.5 Rust-an important compound

A Rust and bust

Rust is an unwanted compound called iron oxide. It forms when the element iron joins with the element oxygen. Rusting can only happen when water is present.

$$iron + oxygen \rightarrow rust$$

Collect

Rust indicator
Nail
Salt water
Sandpaper
Test tube and rack

1 Rusting happens quite slowly, but you can show that it is happening by using rust indicator. This changes colour when iron begins to rust.
   Clean the rust off a nail by rubbing it with sandpaper.
   Put the nail in some salt water.
   Add a few drops of rust indicator and observe.
2 Look at the rusting display.
   All these objects contain the element iron.
   They have all been damaged by rust.
   Crumble some of the rust between your fingers. It is orange and flaky, and not very strong.

1 What happens to rust indicator when rust is present?
2 Choose an object from the display.
   a What is your object and what is it usually used for?
   b How do you know that your object is rusting?
     List your observations.
   c Can the rusty object still be used safely?
     Explain your answer.

CHECKPOINT

B Stop the rot

When iron and steel (which contains iron) rust, the metal loses its strength and shape. It eventually becomes useless. Fortunately there are ways of slowing down rusting, but they are expensive.

1 Clean the four nails with sandpaper.
Use one nail in each of experiments **a**–**c**.

  a Go to the painting area. Dip a nail into paint.
Push the nail into a polystyrene block to dry.
  b Go to the oiling area. Dip a nail into oil.
  c Go to the plastic-coating area. Heat a nail strongly.
Push the hot nail into plastic powder.

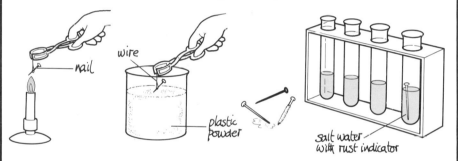

2 Put each nail in salt water. Add a few drops of rust indicator and observe.

1 Why did you have to clean the nails first?
2 Put your results in a table of three columns.
The headings should be *Nail covering*, *Colour of indicator*, and *Did the nail rust?*
3 Explain how paint, oil and plastic coating help to slow down rusting.
4 Give some examples of objects that are protected from rusting by: **a** paint, **b** oil, **c** plastic coating.
5 **Collect** a cartoon sheet. Discuss the best way to protect each metal object from rusting. (You should talk about appearance, cost and how long the protection must last.) Write down the best method of protection in each case.

A Detection

Water is a compound which contains hydrogen. Acids are also compounds that contain hydrogen.

Acids can be dangerous . . .   . . . or quite safe.

Acids can be cancelled out and made safe by compounds called alkalis. Acids and alkalis are detected with indicators.

Peaty soil is acid   Sandy and chalky soil is alkaline

Collect

Safety glasses
Acid
Alkali
2 test tubes
Set of indicators

1 Put a little acid in a test tube.
2 Put a little alkali in another test tube.
3 Add a drop of **one** of the indicators to each tube.
   Note the colour in a table.
   Your table will have three columns.
   The first one should be headed *Name of indicator.*
4 Wash the tubes.
   Repeat with another indicator.

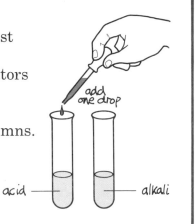

add one drop

acid — alkali

1 What can an indicator be used for?
2 Write down the names of three different acids and describe a use for each of them. (You may use the books in the classroom to help you.)

B In the home

We can use a mixture of indicators to detect acids and alkalis. The mixture is called **universal indicator**. Sometimes it comes in the form of a liquid and sometimes on paper, called **pH paper**. The colours of universal indicator are shown below.

Collect

Spotting tile
Pieces of pH paper
Coloured pencils or crayons

There are a number of household liquids in the room, including tea, coffee, lemonade, bleach, baking powder solution, lime juice, water and vinegar.

Test each one separately for acidity as described below.

1 Put a drop of the liquid on a spotting tile.
2 Add a piece of pH paper.
3 a Write the name of the liquid in your book.
  b Describe the colour that the yellow indicator changes to (use a crayon if you like).
  c Write *acid, alkali* or *neutral* beside each liquid's name.

A Chemical change

Elements and compounds can change into different substances. This type of change is called a **chemical reaction**. Chemical reactions can be very useful.

Chemical reactions can make new substances, like fibre-optic glass

Chemical reactions can give us energy

Collect

Test tube
Test tube holder
Bunsen burner and mat
Splint
Chemicals for one experiment
Safety glasses

Do as many of these reactions as you can.
Decide for each one whether
a a new substance is produced
b there is an energy change.

1 Heat a solid in a test tube.

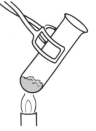

2 Add acid to marble in a test tube.

3 Shake the bottles. Leave them to stand. Shake again.

4 Add zinc to a blue solution. Stir for a few minutes.

5 Mix two solutions in a test tube.

6 Put calcium in water. Test the bubbles of gas produced with a lighted splint.

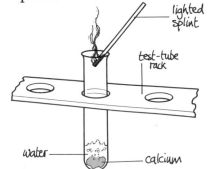

lighted splint

test-tube rack

water

calcium

1 What two changes might happen in a chemical reaction?
2 Write a full report about two of your experiments. Explain how you knew that a chemical reaction was happening.

B Spot the reaction

Discuss the following pictures with your partner.
Decide which show chemical reactions and which do not.
(Are any new substances made? Is there an energy change?)

Write about your decisions.

a
Mixing water and sand

b
Mixing cement

c
Using glue

d
Boiling water

e
Frying an egg

f
Burning garden rubbish

Stick with milk

Milk will react with an acid like vinegar to make a new compound. This compound is a useful glue. Milk glue is usually made like this.

This is how you can test the strength of the glue.

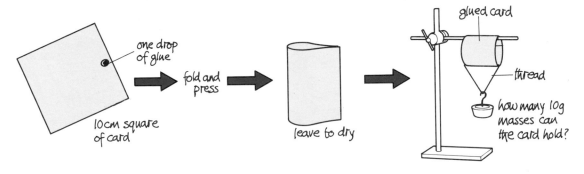

Collect

Milk
Anything else you need

Another way of making glue involves using lemon juice. Your problem is to find the best way of using the lemon juice.

Hints

1 Make the milk glue using milk and vinegar first. Test the strength of it.
2 Make another glue by replacing one of the compounds in the recipe with lemon juice. Test this glue.
3 Continue until you find a strong glue.

Write a report giving your recipe and explaining
a how lemon juice can be used to make glue
b how you tested your glue
c how strong your glue was compared to the original milk glue.

Metals

Metals are very useful

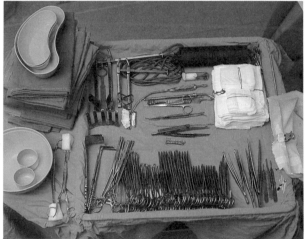

Using the books that your teacher provides, prepare a two-minute talk about the appearance, properties and uses of one of the following metals.

To decide which metal throw a die.

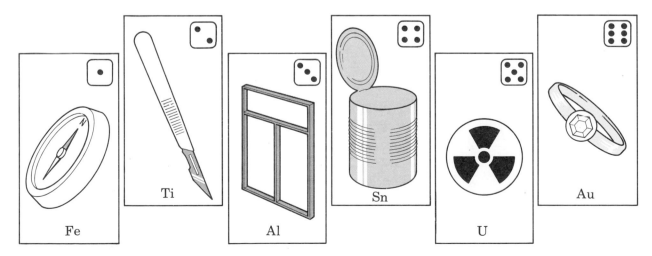

The Earth's resources

A resource is a material which can be made into something useful. The Earth is very rich in resources.

Location	Resource	Raw material
Overground	Air	Oxygen, nitrogen
On the ground	Living things, water, soil	Wool, wood, fuel
Underground	Rocks	Metals, minerals, fossil fuels

The story of how a resource is used is shown in the flow diagram.

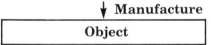
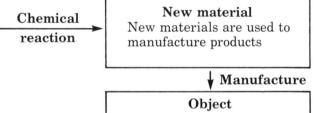

Resource
Useful materials are extracted (taken out) from the air, water, rocks, living things or fossil fuels

↓ **Extraction**

Raw material
Raw materials are then
● used as they are **or**
● changed into other materials by chemical reactions

→ **Chemical reaction** →

New material
New materials are used to manufacture products

↓ **Manufacture**

Object

↓ **Manufacture**

Object

The story of iron can be shown in a similar flow diagram.

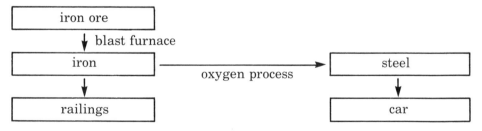

iron ore

↓ blast furnace

iron → oxygen process → steel

↓

railings

↓

car

1 What are the meanings of the words *resource, extracted, manufacture*?
2 Where can useful materials be extracted from?
3 From the flow diagram, describe the story of iron in words.
4 Draw flow diagrams to show how
   a a wooden fence is manufactured
   b a gold ring is manufactured
   c a clay plant pot is manufactured.

3
Good health

A Organs and systems

The human body is made up of millions of cells. Groups of similar cells form **tissues**.

An **organ** is formed from several different types of tissues. It carries out one particular job. When you are fit and healthy all the organ systems of your body work well together.

The cartoon locates your main organs and organ systems.

1
Brain—controls movement and is the centre for thinking, memory and emotions. Part of the **nervous** system.

2
Lungs—supply the body with oxygen. Part of the **breathing** system.

3
Kidneys—keep the blood clean. Part of the **excretory** system.

4
Bones—give support and protection and allow movement. Part of the **skeletal** system.

5
Heart—pumps blood round the body. Part of the **circulatory** system.

6
Stomach—stores and mixes food. Part of the **digestive** system.

7
Sex organs—produce sex cells. Part of the **reproductive** system.

Collect

Body sheet
Scissors
Glue
Stop clock

1 Start the clock. You and your partner have 5 minutes to learn the position, name and job of each organ.
2 Close the book when you have read these instructions.
3 Cut out the organs and the labels.
4 Discuss with your partner where each organ should go. Place the organs on the body and the labels around it. Draw arrows from each label to its organ.
5 Join another group and compare your answers. When you are sure they are right stick the completed body outline into your book.

Discuss each card with your partner.
Use your new knowledge to put the cards into four
sets: cards that describe hazards to your
- digestive system
- breathing system
- nervous system
- circulatory system.

 Copy and complete the table below using your card sets.

Organ system	Health hazards

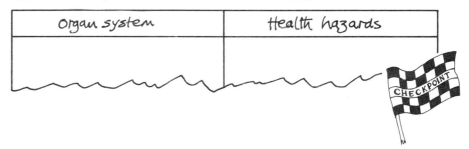

B Avoid health risks

Study the pictures below. Each shows a health risk that some
people will take from time to time.

Wash your hands

Don't slouch

Don't smoke

Don't drink alcohol

 Copy the title of each picture. Write down
the name of the body system or systems
that could be affected.

A Food

You have to eat food for three reasons:
- food gives you energy
- food gives you substances that your body needs for health
- food gives you the building blocks for growth and repair.

A food can contain seven different types of basic food substance.

Water: to carry things round the body and replace lost water.
Fat: to store and provide energy.
Protein: to grow new tissue and repair damage.
Carbohydrate: to provide a supply of energy.
Minerals: to make healthy blood, bones and other tissues.
Vitamins: to take part in important chemical reactions in your body.
Fibre: to keep the digestive system healthy.

Collect

Small pieces of different foods
Clinistix
Albustix
Bottle of iodine
Filter paper
4 test tubes and rack
Stirring rod

1 Stir the food in warm water in a test tube.
2 Divide the liquid into the three remaining test tubes.
3 Test the liquids using tests A–C. Use a solid piece of dry food for test D.

Food tests

A Starch   **B** Sugar   **C** Protein   **D** Fat

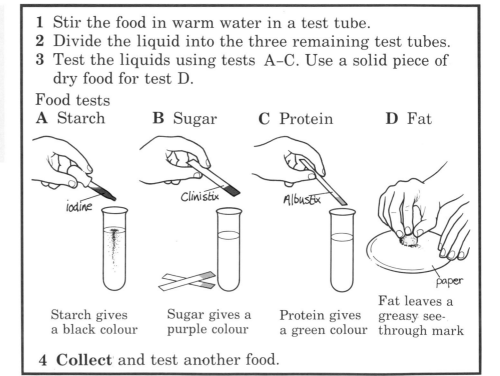

Starch gives a black colour

Sugar gives a purple colour

Protein gives a green colour

Fat leaves a greasy see-through mark

4 **Collect** and test another food.

1 Write a report about your findings.
  Include a short description of each test and your results.
2 Make a table to show what each of the seven basic food substances does.

A balanced diet contains the correct amount of all of the seven basic food substances.

A simple rule is to eat something every day from each of the four food groups shown below. Malnutrition is caused by having a diet that is not balanced.

Milk and milk products

Fruit and vegetables

Bread and cereals

Meat and meat substitutes

1 What is **a** malnutrition, **b** a balanced diet?

2 What four food groups should you eat something from each day to have a balanced diet?
Make up a balanced meal using the foods in the picture above.

CHECKPOINT

B Food sources

Some foods contain a lot of one of the basic food substances.

Lots of carbohydrate

Lots of fibre

Lots of fat

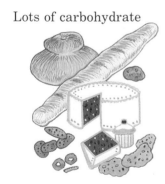

Lots of protein

Lots of vitamins and minerals

Collect a copy of the buffet-table diagram and coloured pencils.

Make and use a colour key to show which foods contain a lot of
- carbohydrate
- fibre
- fat
- protein
- vitamins and minerals.

CHECKPOINT

Healthy teeth

A Why we need them

Human teeth are used to bite food and break it into smaller pieces. There are four types of adult human teeth.

Incisor   Canine   Premolar   Molar

Collect

Mirror
Piece of apple
Graph paper
Jaw diagram

1 Use the mirror to examine your mouth.
   Find and count each type of tooth:
   ● the chisel-shaped **incisors**
   ● the fang-like **canines**
   ● the square, bumpy **premolars** and **molars.**
   Label the four tooth types on your jaw diagram.
2 Eat the piece of apple. Decide which type of tooth does each of these jobs:
   ● scrape and bite the apple
   ● bite and tear the apple
   ● chew the apple.
3 Work with a partner. Use the mirror to count how many filled and missing teeth you have.
   Draw a bar chart to show the number of healthy, filled and missing teeth in your mouth.

Toothache really hurts. It is usually caused by tooth decay which happens like this.

1 Food sticks between and on the teeth.

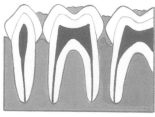

2 Bacteria live and reproduce in the food. This sticky coating of bacteria and food is called **plaque**. Bacteria make **acid** which attacks the outside **enamel** of the tooth.

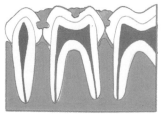

3 The acid eats through the enamel and attacks the soft **dentine** underneath.

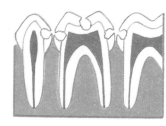

4 The decay reaches the **pulp** in the centre of the tooth where the **nerve** is. Pain begins when infection sets in.

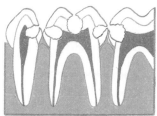

Imagine you are a bacterium living in someone's mouth. Write a short story about how you helped to destroy this person's front tooth and cause toothache. Use all the words shown in **bold** in the passage on tooth decay.

B Toothpaste trial

Collect

Disclosing tablet
Spot of Vaseline
2 squares aluminium foil
Toothpaste
Red pencil
Teeth diagram
Mirror
Microscope

1 Disclosing tablets will stain any plaque in your mouth red. Your teacher will describe how to use them. Use half a tablet each time.
Colour the first teeth drawing red to show where there was plaque.

2 Rub some toothpaste over your teeth. Rinse your mouth out. Test for plaque again.
Colour the second teeth drawing to show where there was plaque.

3 Rub some water over an aluminium square. Then rub some toothpaste over the other aluminium square.

4 Examine both squares of aluminium under the low-power lens of a microscope. Look for differences.

1 Write a report about your experiments. Stick your teeth diagram into your book. Label it.

2 How does toothpaste clean your teeth?

The fate of food

A Digestive system

Food begins its long journey through the digestive system in the mouth. In digestion, large insoluble lumps of food are changed into soluble food which can enter the blood system. The path food takes is shown below.

Mouth—food is mixed with saliva, chewed and swallowed. Digestion starts here.

Gullet—food is passed from the mouth to the stomach.

Fibre—helps food to move through the digestive system.

Carbohydrates, proteins and **fat**—changed to smaller, soluble molecules.

Stomach—food is mixed with digestive juices and stored. Acid kills any germs present.

Small intestine—food is broken down and becomes soluble. Soluble food passes into the blood supply.

Large intestine—water is removed and remaining food is now waste.

Anus—waste food is stored and egested.

Collect

Bingo card
Bag of bingo caller cards

1 Play 'Digestion Bingo' in a group of four or five.
2 One pupil will be the caller. The caller takes a card from the bag and reads the sentence on it.
3 If a player's card has the part of the digestive system described, then that part can be covered or crossed off.
4 The winner is the first player whose card is completely covered or crossed off.

1 Why must food be made soluble by digestion?
2 Imagine you are a slice of bread containing starch and fibre. Describe your journey through the digestive system. Say what happens to the starch and fibre.

Digestion is a complicated set of processes. When things are difficult to understand it is a good idea to use a model to help explain what is going on. You will make a model of the digestive system in the next experiment and use it to study what happens during digestion.

Bread contains a lot of starch which is **insoluble**. Starch is changed into **soluble** glucose in the digestive system.

1 Set up the experiment shown below.
2 Wash the outside of the bag, then put it in the test tube of water.
3 Test the water for starch and glucose at the start of the experiment and after 20 minutes.

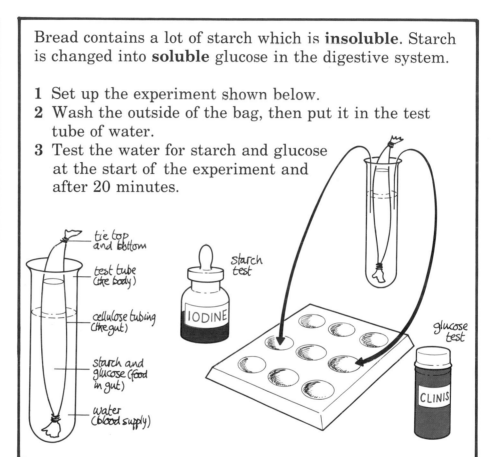

Write a short report about the experiment. Include a
- title
- labelled drawing
- description of what you did
- sentence describing the result
- sentence describing how the result explains what happens in digestion.

B Digestion summary

Collect a summary sheet. Cut out the boxes. Make up a flow diagram to show the fate of food in digestion.

A How they operate

One of the most common illnesses in Britain is heart disease. The human heart is made mainly of muscle. It weighs about 375 g and is about the size of a man's fist.

Your heart beats roughly 100 000 times every day.
Blood enters the heart from veins. A heart beat pumps it out at high pressure through the arteries. In a lifetime the heart will pump 300 million litres of blood. You can feel the pumping action of the heart in your arteries. This is called a **pulse**.

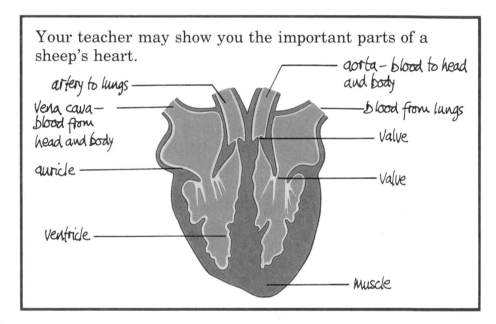

Your teacher may show you the important parts of a sheep's heart.

- artery to lungs
- Vena cava – blood from head and body
- auricle
- ventricle
- aorta – blood to head and body
- blood from lungs
- Valve
- Valve
- Muscle

1 Describe the sheep's heart.
2 **Collect** a heart diagram. Label the parts. Complete the description of how it works by adding the missing words.
3 Write two sentences about the work that the human heart will do in a lifetime.
4 What is a pulse?

You breathe air into your lungs. If you hold your ribs as you breathe in you can feel your lungs filling with air.

The lungs pass oxygen from the air into the blood supply. Oxygen is then delivered to all your cells to keep them alive. Healthy lungs are good at taking oxygen out of the air. Unhealthy lungs are not.

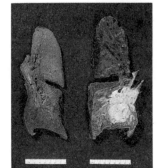

Healthy lungs   Unhealthy lungs

1 What pulse readings can be used as a sign of fitness?
2 Write a few sentences to describe the advantages of having a healthy heart.
3 Why is it important to calculate your exercise range before you try to improve your fitness?
4 Complete your personal training graph.

Training will also improve your breathing system. The average lung volume of an adult is 5 litres. During exercise breathing becomes deeper and faster. A top athlete is able to breathe a larger volume of air in and out than an unfit person. Even when you breathe out as hard as you can some air is left in your lungs.

1 Use the apparatus for measuring the volume of air that you can breathe out. Take a deep breath. Breathe out as much air as possible through the rubber tubing.
2 Record the volume of air that you breathed out.

1 Draw a diagram to show how you measured the biggest volume of air that you could breathe out.
2 If you were training an athlete how could you check that your training plan was working?

B Training plan

Make up a one-week training scheme for yourself or a friend. Write down
- suggestions for the type of exercise
- the exercise range (range of pulse rates that should be aimed for)
- how an improvement in fitness could be measured.

A Smoking and drinking

Tobacco smoke contains chemicals that can damage your lungs.

Tar and other substances in tobacco smoke irritate the lungs. You cough more, and more germs get into your lungs.

Nicotine is a poison. It makes the arteries narrower. Your heart then has to pump harder to push blood along.

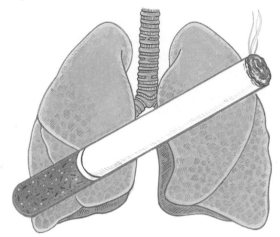

Carbon monoxide gas is quickly picked up by the blood. It prevents the blood from picking up oxygen, and so you have to breathe faster and your heart has to beat faster.

Smokers may feel that they need nicotine. They have become **addicted**.

Your teacher will demonstrate a smoking machine. This machine catches some of the harmful chemicals in cigarette smoke.

Watch the appearance of the wool and the colour of the universal indicator. Smell the glass tube.

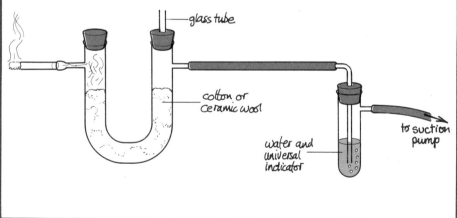

Imagine that you are a doctor.
Write a full report about this experiment. Your report should include a diagram and a description of what was done. It should describe your observations.
Also include your opinion about whether cigarette smoke should be going into people's lungs.

Alcohol is a drug. You can become addicted to it by drinking it regularly, even in small amounts. A heavy drinker runs the risk of serious liver damage. Drinking too much alcohol in one go can kill you. Alcohol has an effect on behaviour.

Here are two important facts about alcohol to learn.
● All of these drinks contain about the **same** amount of alcohol. This is called **one unit**.

1 unit
 ½ pint of beer
 glass of wine
 glass of sherry
 small whisky
 small vodka
 small martini

● It takes your liver about one hour to remove one unit of alcohol from the blood.

1 Write a short story to describe **one** danger of drinking alcohol.
2 Give the name and volume of four drinks that each contain one unit of alcohol.
3 Copy and complete this table, adding your own examples.

Alcohol drunk	Units of alcohol	Time for liver to remove alcohol
Jack drank: 3 pints beer 2 whiskies		
Jill drank: 2 vodkas 1 pint beer		

CHECKPOINT

B Advertising

Collect
Advertisement
Felt pens
Glue
Paper
Stencils

Advertisements for cigarettes and alcohol usually connect smoking and drinking with something good in life. You know how dangerous they can be. Change and reword your advert so that it warns people about the dangers of smoking or drinking.

CHECKPOINT

A Brain work

The human brain is truly amazing. It weighs about 1.3 kg. It can think faster than the most expensive computer. It can do far more jobs than any machine.

Even more importantly, it can learn new facts and skills. The brain is made up of millions of tiny nerve cells. These can pass messages to each other. They work together to do all sorts of important jobs.

1 Thinking, memory, emotion
2 Hearing
3 Sight
4 Movement
5 Co-ordination
6 Breathing, digestion, heart rate

front   back

Work with a partner in the next two activities.

Collect

Star drawing apparatus
4 star shapes
Stop clock
Mirror

Memory test

Imagine that you have just witnessed a bank robbery. There is a picture of the robber on the classroom wall. Stand about 1 metre from it. Your partner will uncover the picture for 10 seconds only. Write a full description of the robber in the picture.

Learning test

1 Sit on a stool. Adjust the mirror so that you can see the star drawing.

2 Look in the mirror only (no cheating!).
Time how long it takes to draw round the star keeping inside the lines. Repeat this with the other three star shapes.

1 Write the best description you can of the human brain. Use the information and drawing on the opposite page.
2 What jobs must your brain be doing now, as you answer these questions?
3 Describe the memory test. What was accurate about your description of the robber? What was inaccurate?
4 Write a report of the learning test. Include a table of results with two columns. Did you improve with practice?

B Ways of learning

Sometimes you have to remember facts at school.
Your memory can be improved if you sort facts into groups.

You have 30 seconds to do each of the tasks described below. After each task go back to your seat and write down the names of as many objects as you can.

1 Go to tray 1.
   Look at it for 30 seconds only.

2 Go to tray 2.
   Arrange the objects into four groups in 30 seconds.

---

1 Is your first or second list most correct?
   Did sorting things into sets help you to remember them?
2 The cartoons show some different ways of learning.

a Doing

b Observing

c Listening

Give an example of something you have learned in each of these ways.

Design a game

Your problem is to design a board game which
- is fun to play
- teaches the players about healthy living

Hints
- Your group should decide what type of game to produce.
  Here are three suggestions:

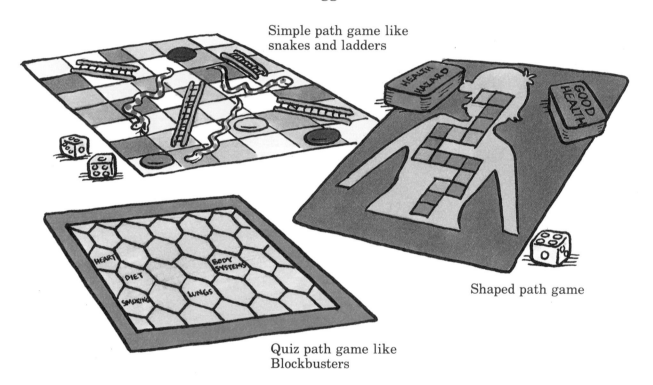

Simple path game like
snakes and ladders

Shaped path game

Quiz path game like
Blockbusters

- Decide on the aim of the game and the rules.
  (Games with simple rules usually work best.)
- Make a list of at least five **good health** points and at least
  five **bad health** points.
  You must use these in your game.

Collect

Blank game board
Stencils
Coloured pencils
Coloured paper
Scissors
Dice
Plasticine

1 Design and make your game.
2 Play your game to make sure that it works.
3 Try out some of the other games made in the class.

Talkabout

Smoking and health

How would you persuade someone to stop smoking?
Some of the ideas that you can use are given below.
You will be able to think of others.

Some survey information
A survey of more than 10 000 11–16 year olds produced these results.

Where do smokers smoke?

Should children be allowed to smoke at home?

What percentage of boys and girls started to smoke in their first year at secondary school?

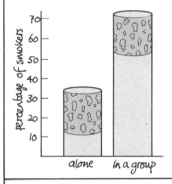

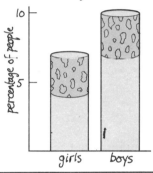

Some medical facts

9 OUT OF EVERY 10 PEOPLE WHO GET LUNG CANCER ARE SMOKERS

IF YOU SMOKE YOU'RE TWICE AS LIKELY TO HAVE A HEART ATTACK

A PREGNANT WOMAN WHO SMOKES MAY HAVE A SMALLER BABY THAN SHE OTHERWISE WOULD HAVE

HOT SMOKE DAMAGES THE CELLS THAT LET YOU SMELL AND TASTE

Some advice

1 Work in a small group.
 Discuss the ideas shown above, and any of your own.
2 Write a script for a short play or a radio advertisement to encourage people to stop smoking.
3 **Collect** a tape recorder and a blank tape. Act out your script.
4 Play your tape to the class.

Body defences

Even when you are fit you can wake up in the morning feeling unwell. Very often it is because you have some sort of infection. Infections can be caused by bacteria, fungi and viruses. These are all types of microbe.

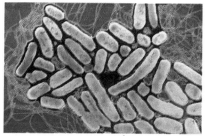

Food poisoning can be caused by salmonella bacteria [Magnified 3000 times]

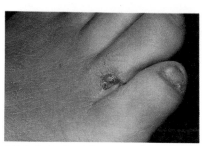

Athlete's foot is caused by a fungus

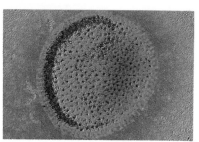

Flu is caused by a virus [Magnified 350 000 times]

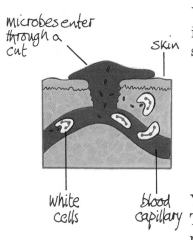

microbes enter through a cut

Skin

white cells

blood capillary

Your body has a defence system that can fight off most infections. Your skin and blood are part of this defence system.

● Skin prevents microbes from entering your body. When this barrier is broken, blood quickly clots to seal the wound with a scab.
● Blood contains red and white cells. Red cells carry oxygen to all parts of the body. It's the job of the white cells to attack, digest and destroy invading microbes.

Your blood also carries special proteins called antibodies. These recognise microbes in the blood and destroy them or make them easier for the white cells to find.

1 Make a table showing:
   ● three types of microbes
   ● descriptions of the infections they cause.
2 Copy and complete the flow diagram below.

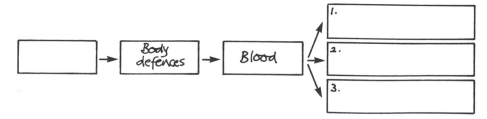

Body defences → Blood

1.

2.

3.

4

Everyday forces

Describing forces

A Shaping forces

Forces affect our lives every day.

Force the door

A force 9 gale is expected

Force out the last drop

It's protected by a force field

Forces cause changes. Forces can move or stop an object. They can change the direction of a moving object. Forces can also change the shape of an object.

Collect

Plasticine

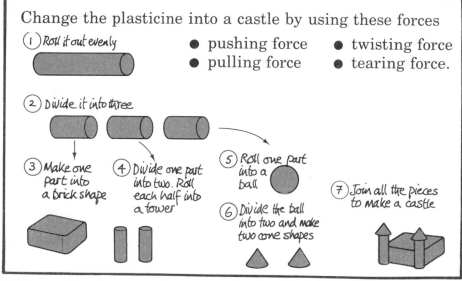

Change the plasticine into a castle by using these forces

- pushing force   - twisting force
- pulling force   - tearing force.

① Roll it out evenly

② Divide it into three

③ Make one part into a brick shape

④ Divide one part into two. Roll each half into a tower

⑤ Roll one part into a ball

⑥ Divide the ball into two and make two cone shapes

⑦ Join all the pieces to make a castle

1 Copy the diagram into your book. Under each step write down the types of force you used to make that shape.
2 Make a list of things that can happen to a moving object when a force acts on it.

You use forces all the time.
Collect a tray of objects. Try out each of the everyday activities shown below.
Discuss whether you are using a pushing, pulling, tearing or twisting force.

Open the jar

Pick up the pen

Crumple the paper

Take off the lid

Write a poem

Put it in the bin

Make a table to record your discussions. The table should have two columns.

B Forces everywhere

Forces are in action everywhere. Think about the pictures below. Work out what types of forces are shown.

Lock the door

Just weight

Turn the page

Horse and cart

Copy the title of each of the pictures.
Describe what you see. Write down the type of force being used.

Forces at a distance

A Non-contact forces

Contact forces must touch an object before they can make something happen. Pulling, pushing and twisting are contact forces. There are other forces which can work from a distance. Three examples of **non-contact** forces are gravity, electricity and magnetism.

Play the three games. They will help you to understand these non-contact forces.
Work with a partner.

Collect

Long cardboard tube
Retort stand
Ruler
Plasticine

Game 1: Bat the Ball
The Gravity Game

An object falls because the force of gravity pulls it towards the centre of the earth.
1 Make a small ball from plasticine.
2 Your partner will drop the ball down the tube.
   Try to hit the ball as it escapes from the tube. Try three times.
3 Now let your partner try.

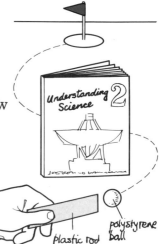

Collect

Polystyrene ball
Piece of cloth
1 plastic rod
Timer
Piece of filter paper

Game 2: Crazy Golf
The Electric Game

Rubbing a plastic rod gives it an electrostatic charge. A charged rod can attract some objects.
1 Rub the rod with the cloth to give it an electrostatic charge. Time how long it takes to get the ball in the hole. You must not touch the ball with the rod.
2 Now see if your partner can beat your time.

Collect

Petri dish
Face outline
Small magnet
Iron filings
Timer

Game 3: Make a Face
The Magnetic Game

Iron is attracted to a magnet.

1 Complete the face by adding two eyebrows, the middle bits of both eyes and by blacking out one tooth. Time how long this takes you.
2 Now see if your partner can beat your time.

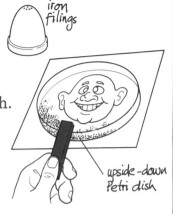

iron filings

upside-down Petri dish

1 Write down three contact forces and three non-contact forces.
2 Name the non-contact force used in each game and describe how it was used.
3 Copy and complete the table below:

Game	Force acts on...	Object attracted to...
1		The centre of the earth

CHECKPOINT

B Using non-contact forces

Forces that act from a distance are used in everyday objects. Look at these pictures.

a b c d

Write down the non-contact force being used in each picture.

CHECKPOINT

A Friction

A small push can move a car. So why does it stop again? When any object moves over a surface the force of friction slows its movement. For example, with a car, friction is caused when the tyres rub along the ground.

Think about trying to push a heavy box along the ground. If your pushing force is less than the frictional force, then the box will not move. If your pushing force is equal to the frictional force, the box will move at a steady speed. If your pushing force is greater than the frictional force the box will move faster and faster (accelerate).

Ways of reducing friction

You can use an elastic band and coin to investigate friction. If the elastic band is pulled back the same distance each time the coin will get the same pushing force each time.

elastic band

3cm only

Discuss how you can make the coin move further along the bench **without** giving it a bigger push.
You will have to think of ways of reducing the force of friction between the coin and the bench.
The pictures on this page will give you some clues.

> **1** Decide
> - how you will measure how far the coin goes
> - how you will make the frictional force smaller
> - how you will make your experiments fair.
> **2** Everybody in the group should write down their ideas.
> **3** Agree on the best ideas. Write down what you need.
> **4** **Collect** the equipment you need.
>   Do as many experiments as you can.

1 How do you know when friction is acting?
2 What forces are balanced when a car is moving at a constant speed?
3 Why does a bike stop when you don't pedal, even if you do not use the brakes?
4 Write a short report on the coin experiments.
   Include a drawing with labels to show how you carried it out. Remember to write down what you did to make the experiments fair.

B Using friction

Friction can be very useful. Look at these examples of how friction is put to good use.

Choose three of the pictures. Describe what is happening in each. Use the word *friction* every time.

A Getting the right balance

Here is a brain teaser.

There are 15 pupils in a class. 14 of them have exactly the same weight. One pupil is 2 kg heavier than the others. How could you use a see-saw to find the odd one out?
You can only use the see-saw three times.

Collect

See-saw
14 coins **or** metal discs
1 larger coin **or** metal disc
Ball of plasticine
Large beaker of water

1 Discuss and then solve the brain teaser with a partner. (There is a hint on how to start at the bottom of the next page.) Try your solution with this equipment.

2 Here is another puzzle to solve.

How can you make a lump of plasticine float on water?

When an object floats there must be balanced forces. The downwards push of the plasticine is balanced by the upwards push of the water.
Try to float the plasticine on the water.
You can make it into any shape you like, but you are not allowed to throw any of it away.

DESIGN DRAWING

1 Describe how you solved the see-saw brain teaser.
2 When the see-saw is level what must be true about the forces on each side?
3 Make a drawing to show how you solved the floating problem. Describe the shape of plasticine that worked best.
4 Why does a beach ball float on water?
  Use your understanding of forces to answer this question.

B Losing your balance

Look at the picture puzzles below. In one cartoon of each pair, the forces acting are balanced. In the other they are not.

Accelerate   Walk tall   Let fly

Copy the title of each cartoon pair.
Explain what is happening in each picture using the idea of balanced and unbalanced forces.

Puzzle hint: Start with seven people on each side. If the see-saw balances then the person left out is heavier. If not, the heavy person will be on the side that goes down . . .

A Moving joints

Bones move against each other at joints. Your muscles provide the force needed to move the bones.

1 Sit on a chair.
2 Put your hand on the muscle at the back of your thigh. Slowly swing your leg back and forth below your chair.

3 Now put your hand on the front of your thigh. Swing your leg again.

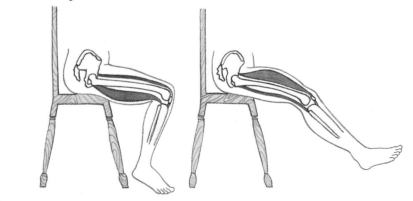

1 How many leg muscles are needed to bend the lower leg up and down?
2 What happens to the muscle at the back of your thigh when you raise your lower leg?
3 What type of force is this muscle providing?
4 Describe what happens to the muscles when you straighten your leg.
5 What happens to the shape of the muscle at the back of your thigh when you swing your leg backwards?

B Mighty muscles

Movement of your body depends on two muscle sets at each joint. While one muscle contracts (shortens) to pull on a bone, the other muscle relaxes. Muscles always work in pairs like this because they can only provide a pulling force. They never push.

1 Roll up one of your sleeves.
2 Show off your muscles like a body builder.
3 Watch what happens to the muscle at the front of your arm.
4 Now make your arm as straight as possible. Look at the muscles at the back of your arm.

 1 Copy and complete these drawings to show what happens to the muscles.

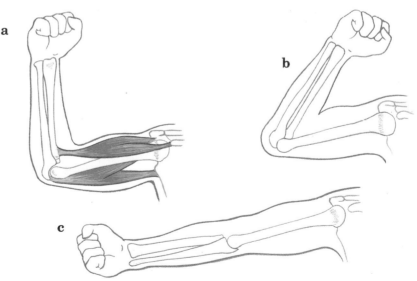

a

b

c

2 Why must muscles always work in pairs?

73

Choosing materials

A Testing properties

Forces can damage objects. When anything new is designed it must be able to stand up to the types of forces it will meet. The most suitable materials must be used.

It's hard to predict the force of storm winds

Some of Britain's roads are breaking up because of the very heavy traffic on them

Architects, engineers and designers choose materials because of their properties. For example, steel is used for building bridges because it is hard, strong and does not bend easily. Steel can withstand pushing, pulling, twisting and tearing forces.

Paper and plastic are used to make carrier bags. These materials must have certain properties. They need to be
- flexible for easy storage
- strong enough to carry a load without tearing
- hard wearing for long life.

Collect

1 carrier bag
Box of 100 g masses
Mass carrier
Wooden block
1 kg mass
Coarse sandpaper

Organise a class competition to find the best carrier bag.

Carry out these three tests on your bag.

1 Which bag is easiest to store away? Test the flexibility by folding your bag as small as possible.
2 Does your bag tear easily? The diagram shows how to do this test. Add the 100 g masses one at a time.

3 Does your bag wear well? Rub your bag 50 times with the coarse sandpaper wrapped around the wooden block. The 1 kg mass should help to make the force the same each time. Do the tear test again on the part of the bag that you rubbed.

1 Copy and complete the table to report the results of the tests you did on your carrier bag.

Comparing carrier bags		
Property tested	Test method	Result
1. Strength		
2.		
3.		

2 Compare your results with the rest of the class. What are the two best bags?

3 Do you think your comparison was fair? How could it be improved?

B More choices

The pictures below show some examples of materials chosen for a certain job.

Bricks for houses

Wooden hockey stick

Metal drill bit

Plastic floor tile

Copy the title of each picture. Under each title write down what properties of the material are important for this use. Write down the types of forces that the object must withstand.

Consumer report

When you want to buy something you need to be sure that it will work well. There are many magazines giving advice on what to buy; there may be some in the classroom to look at.

These magazines compare different makes of a product. The product is tested in fair experiments. Each is usually repeated many times so that we can trust the results.

Your task is to test three makes of one product and to prepare a report like the ones in the magazines. (These are often called consumer reports.)

Carry out tests on one of these products:
- three makes of **clear sticky tape**
- three makes of **paper clip**
- three makes of **clothes peg**

1 Work in a group of about four people. Discuss what you want to test. One thing you **must** test is **strength**. **Collect** the information sheet to show how this can be done.

2 Decide what other test(s) you want to carry out. Write down a list of equipment you need.

3 **Collect** the equipment you need. Carry out your tests. Keep an accurate record of what you did. Keep an accurate record of your results.

You are going to present a report to the rest of the class. Each member of the group will speak.
- *First person:* you will say **what** tests your group carried out and explain why you thought that these were the most important tests to do.
- *Second person:* you will describe **how** you carried out each test (your methods) and explain how you tried to make your experiments fair.
- *Third person:* you will present your **results**. Use tables and graphs to make these results easier to understand.
- *Fourth person:* you will **sum up**. Tell the rest of the class which make of the product they should buy.

Prepare a written summary of what **you** are going to say.

Talkabout

Design flaws

What design flaws can you spot in each of these pictures?

Think about
- the sort of forces that are acting
- how the design could be improved
- some examples of well designed objects.

New materials

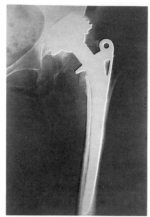

Artificial hip joint

The real thing

Our skeleton has 187 joints. Our joints are strained all through life. As we get older they move less smoothly. There is more friction between the bones and we slow down. Diseases like arthritis can make things worse. In some cases the joint becomes so damaged that any movement is painful. If this happens to the hip joint walking becomes very difficult.

A successful artificial hip joint has been designed. Every year many people who are crippled by arthritis have operations to replace their hip joints. They can then walk again.

An artificial joint has to have similar properties to a natural one. It must allow movement in the same directions. It must be lightweight, but very strong. It must have smooth surfaces. It must not break up or rot in the body. It must not be made of materials that could harm the body.

Plastic and the metal titanium are an ideal combination. Titanium is light and strong, and it does not rot. It has a shiny, smooth surface. In a hip joint the titanium ball fits neatly into a plastic socket. This plastic does not wear or rot. The shape of the parts allows movement in all directions. The same materials are being used in the design of other important joints, such as knees.

1 What two materials are mentioned in this passage?
2 What are they used to build?
3 What types of forces do you think act on the hip joint as you move? (Think about the sort of movement you have at your hip.)
4 Copy and complete the table below.

Materials in an artificial hip joint

Material	Property of material	Why this is important
Titanium	1.	
	2.	
	3.	
	4.	
Plastic	1.	
	2.	

5
Tiny ideas

A Begin with ideas

People are very good at thinking. Some animals also seem to be good at thinking. A cat can find food in the garden and a dog knows how to frighten the postman. However, people can find answers to new and tricky problems. Look at the puzzle below. You can work it out very quickly, but try showing it to a gerbil or a budgie!

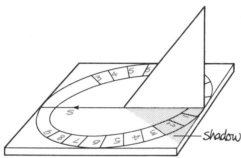

What is this used for?

Which direction should the line marked S point in?

How would you read the scale?

If you want to build this instrument then ask for the instruction sheet

Science helps people to think. Thinking will help you to explain things.

You have a **problem** → You do some **thinking** → You work out an **explanation** → Now you can take **action**

Collect

Activity sheet
What you need for
one activity

Do some (or all) of these activities. The instructions are on the report sheet for each activity.
For each activity you will have to describe what you see. Then you will have to think of an explanation for it.

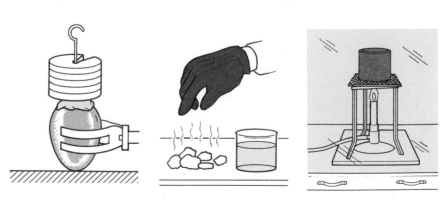

1 Crack an egg   2 Dry ice   3 Can opener

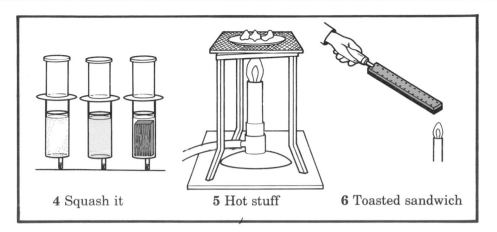

4 Squash it **5** Hot stuff **6** Toasted sandwich

For each activity complete the report by
- describing what you saw (your **observations**)
- trying to explain what you saw (your **ideas**)

Here is an example.

Report on activity 9

Name of activity: *Air freshener*

Method
Place the air freshener on a table and open it.

Observations
A smell of Norwegian wood filled the air in the classroom

Ideas
There is a small forest growing inside the plastic container.

B Share your ideas

Collect

Poster paper
Coloured pencils

Your group has to make a poster to explain one of the activities. You have 30 minutes to make the poster.

What to do
- Each person in the group will read their **observations** aloud. There should be no comment from the other members of the group while this is being done.
  Everyone in the group then agrees which observations to show on the poster. Write these observations on the poster.
- Each person then reads aloud their **ideas** for explaining the observation.
- The group now has 10 minutes to talk about and then to agree on the best ideas.
  Write these ideas onto the poster.

A Practical clues

It is sometimes useful if you can find a pattern when you are thinking about things that are new to you.

34	32	28	24	20
1 A	B	C	D	E

(row 2: pie-segment circles) 2 A B C D E

(row 3: shapes — square, triangle, hexagon, pentagon, diamond) 3 A B C D E

(row 4: dominoes) 4 A B C D E

(row 5: matchstick arrangements) 5 A B C D E

Which is the odd one out in each set? (Answers at the bottom of the page)

A pattern will help you to
- collect the observations into sets
- find explanations more quickly.

Find a pattern in each of these practical exercises.

Fingerprints

Collect

Set of fingerprint halves

Match eight half pieces to make four complete fingerprints, like this example. Stick the fingerprints into your book.

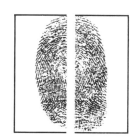

The magic strip

Collect

Thin bimetallic strip
Bunsen burner and mat

1 Look at the bimetallic strip. It is made of two different metals.
2 Heat one of the metal sides of the strip.
   Cool the strip under a tap.
3 Turn the strip and heat the other side.
4 Look for a pattern in the final shape that the strip has both times.

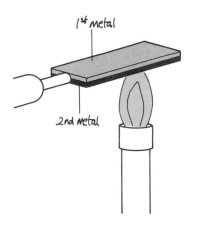

Puzzle answers: 1 A, 2 C, 3 E, 4 E, 5 A

1 What happens to the bimetallic strip when the first metal side is heated?
What happens when the other metal side is heated?
2 Draw the shape of the strip when
  a the first side is heated
  b the second side is heated.
  (Colour the two sides of the strip differently.)
3 What does the strip do when it begins to cool?
Draw what might happen if the strip is cooled with ice.
(Test the strip to see if you were right.)

B Flame colours

A pattern is sometimes clear from the results of a number of experiments. A pattern can help you to work out what will happen in a future experiment.

Collect

Set of solids
Test rod
Bunsen burner and mat

1 Light the bunsen burner.
  Use a blue flame.
2 Dip the rod into one of the solids.
  Hold it in the flame.
3 Note the name of the chemical
  and the colour of the flame.
4 Clean the rod in water.
  Repeat for all the solids.
  Look for a pattern in the results.

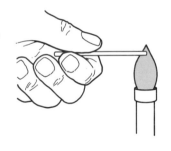

1 What do these solids do to a blue bunsen flame?
2 What colour would you expect if the following solids were put in a flame?

Copper Sulphate

Sodium Sulphate

Try them and see if you were right.

Bits and pieces

A Break up

In the following experiments you are going to
- make observations
- find a pattern to these observations
- try to explain the pattern.

Collect

Purple crystal
5 test tubes
Beaker of water

Experiment 1

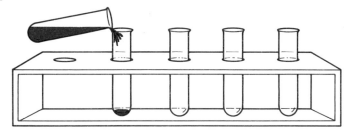

1 Add **one** purple crystal to a test tube of water. Mix until the crystal dissolves.
2 Pour one-fifth of the coloured liquid into a second test tube. Fill with water and mix.
3 Pour one-fifth of the contents of this tube into a third tube filled with water.
4 Continue until there seems to be no colour left in the liquid.

Collect

Petri dish
White crystal
Blue crystal
Spatula

Experiment 2

1 Fill a petri dish with water.
  Put it on the table. It must remain steady.
2 Gently put a white crystal into the water at one end of the dish.
3 Gently put a blue crystal into the water at the other end of the dish.
4 Wait for a few minutes.

1 Make observations

    a How many purple crystals did you put into the first tube?

    b Was there any purple substance in the second tube?

    c Draw a diagram of your results in experiment 1 to show how many tubes contained the purple substance at the end.

    d Draw a diagram to show what happens after a few minutes in experiment 2.

2 Find a pattern

Discuss the experiments with your partner and try to find a pattern. Describe what crystals do in water. Use words like *crystal, spread, colour, becomes invisible* and *still there*.

3 Likely explanations

Discuss the pattern that you find.

What do you think causes this pattern?

Use words like *pieces, tiny* and *break up*.

B Tiny pieces

Collect

5 test tubes
Test tube rack
Mustard powder
Wooden splint

This experiment will help to explain how a substance can spread if it is made of tiny pieces.

1 Add a tiny amount of mustard powder to a test tube of water. (Use just enough to cover the end of a wooden splint.)

2 Follow the instructions for experiment 1 on page 84.

1 Write a title for your experiment. Describe what you did.

2 Copy the diagram below.

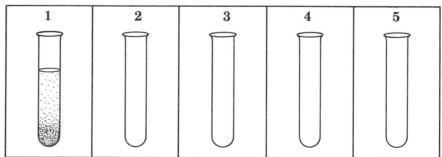

Add dots to your diagram to represent the number of mustard grains that you can see in each test tube.

3 Write down a likely explanation of why mustard can spread so far.

A Use the clues

It is sometimes difficult to explain an observation. You have to think about it for a long time. You may work out different ways of explaining it. Eventually you will decide on a **likely** explanation.

● Ozone layer depleted?   ● Santa on holiday?
● Christmas in Australia?   ● North pole melts?

Collect

Clue cards for the game of Inspector Cluedo

In the following game you have to study a set of cards. Each card contains a clue about the terrible murder that happened in the farmyard.
Your group has to work out a likely explanation which answers **all** the following questions:
  ● Who was murdered?
  ● What weapon was used?
  ● Where did the murder take place?
  ● When did it take place?
  ● What was the motive for the murder?
  ● Who did the terrible deed?
Hint: Look for a pattern in the clues.

You must **not** let another group hear any of your ideas.

1 Write down a likely explanation of the murder at the farmyard which answers all the questions.
   Give reasons if you can for each of your answers.
2 Look at the **right answer** card in the envelope.
   What is the right answer?
   Why is this answer surprising?
3 You can give a good explanation if you find a pattern in the clues.
   a What does the word *pattern* mean here?
   b Why is it useful to find a pattern in the clues?

B Particles

When you dissolve sugar in tea the sugar seems to disappear. However, it must still be there as the tea tastes sweet. You tried to explain this kind of observation in Topic 5.3. One likely explanation is that the sugar breaks up into tiny bits called **particles**. These bits are too small to see.

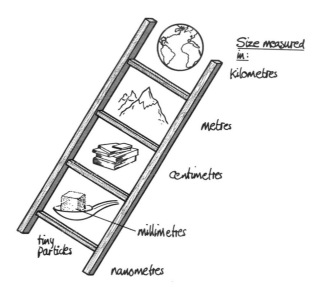

1 What happens to the size of an object as you climb up the ladder of size?
2 Copy the ladder into your book. Discuss which rung a person should be shown on.
3 Explain why you cannot see one particle of sugar.

Collect

Clean container
Bunsen burner and mat
Tripod stand
Bottled water
Safety glasses

1 Inspect the container. Make sure that it is clean.
2 Pour some bottled water into the container and heat it as shown below.

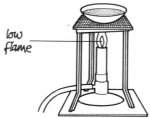

low flame

3 The liquid will eventually disappear. Look at the container when it is cool.
Explain what you see. Use the words *tiny particles* in your explanation.

Moving particles

A Smells

Smells get around. A male emperor moth can smell a female at a distance of 11 km. Your nose is also an excellent smell detector. It can detect over 3000 smelly substances.

> **1** Your teacher will place a dish of ammonia solution in the room.
> **2** Move slowly towards the dish. Find out where you can detect the smell.

1 Copy the plan below.

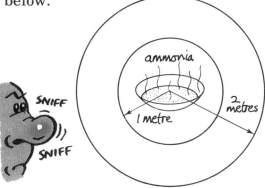

2 Add dots to your plan to show how much ammonia was present at different distances.
Hint: Use a lot of dots for a strong smell.
3 What must ammonia be able to do to spread through a room?
4 Describe the nicest smell that you know.
Draw a picture to show how this smell can spread around a room. Use dots to represent particles.

B Brownian motion

In 1827, a Scottish scientist called Robert Brown discovered something that amazed him. When he looked at pollen grains under a microscope they looked as if they were dancing!

This kind of movement is now called **Brownian motion**. You can also observe Brownian motion.

Collect

Microscope
Bench lamp
Slide
Cover slip

1 Use a high-power magnification.
2 Take the slide to the hot milk in the room.
3 Put one drop of hot milk on the slide and cover it with a cover slip.
4 **Quickly** return to the microscope. Focus the microscope on the slide.

1 Describe what you did.
2 Describe what you saw. Use words like *droplets, shaped like, moving, quick/slow, direction* and *jiggle*. Draw a diagram if it will help your description.
3 Your teacher may show you how smoke looks under a microscope.

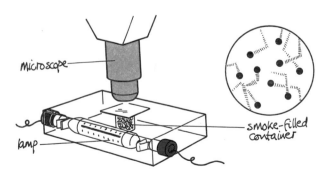

Draw and describe what you see.
4 What explanations can you think of for Brownian motion? Make a list of your ideas.
Then discuss these in class with your teacher.

Spaced out particles

A In between

You can add sugar to a cup which is already full of tea. The tea will not slop over the side of the cup. Why is this? Where does the sugar go?

There are spaces between the particles of the water. The sugar fits into these spaces. Most substances are made of particles that have spaces between them.

Collect

Dish of jelly
Tweezers
Borer
Dropper
Bottle of acid
Bottle of alkali

1 The jelly has universal indicator in it. Use the borer to cut two small holes in the jelly. Pull the bits out with tweezers.
2 Fill one hole with acid and the other with alkali, as shown in the diagram.

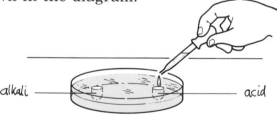

3 Leave the dish **without touching it** for about 10 minutes.

1 Explain why a full cup of tea does not slop over when sugar is added carefully to it.
2 Draw and describe the result of your experiment.
3 Why was universal indicator added to the jelly?
4 Write down a likely explanation for the result of your experiment. Use words like *particles*, *move* and *spaces between*.

B Mysterious spaces

Look at the photograph of stars in part of our galaxy. There is space between the stars. Scientists think that there is also space between particles.

1 Your teacher is going to add exactly 100 cm³ of water to exactly 100 cm³ of alcohol.
  Write down the result that you **expect** when the two are added together.
2 Write down the **measured** amount when the two are added together.
3 Repeat instructions 1 and 2, but add 100 cm³ of rice to 100 cm³ of peas, instead of water to alcohol.

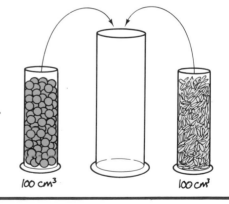

100 cm³   100 cm³

1 Put your results in a table with the headings *Mixture*, *Expected result* and *Actual result*.
2 Draw a diagram to show what happened to the rice when it was added to the peas.
3 Explain the results of both experiments. Use words like *particles* and *spaces*.

CHECKPOINT

91

Making a hypothesis

A About the moon

When you think of a likely explanation for an observation then you are making a **hypothesis**. You can test out your ideas by doing experiments and examining your results.

For example, Sian and Nora observe that the moon's appearance changes during a month.

3rd day of month	7th day	14th day	21st day
1st observation	2nd observation	3rd observation	4th observation

Sian's hypothesis:

DIFFERENT PARTS OF THE MOON ARE LIT UP BY THE SUN AS THE MONTH GOES BY

Nora's hypothesis:

DIFFERENT PARTS OF THE MOON ARE HIDDEN BY THE SUN AS THE YEAR GOES BY

1 What is a hypothesis and how should you test it?
2 Sian and Nora decide to make some more observations of the moon for the next month. This is what they see.

day 30   day 38   day 42   day 52

a Which girl has the best hypothesis?
b There is a pattern in the observations. What is it?
c Explain the pattern using words like *phases of the moon, every 28 days, month, goes round* and *sunlight*. Reference books and the experiment below will also help.

Collect
Light-dark ball

1 Look towards a window.
2 Hold the light-dark ball towards the window.
  The light half should face the window. You should only be able to see the dark half.
3 Move the ball slowly around your head, always keeping the light half facing the window. How does the lighted shape appear to change?

CHECKPOINT

B About particles

We now have several hypotheses about particles.

● Everything is made of particles.

● Particles are moving.

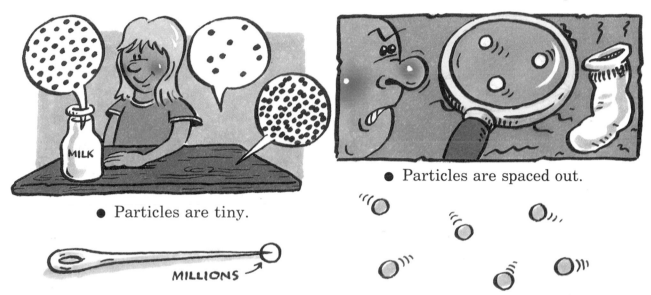

● Particles are tiny.

MILLIONS

● Particles are spaced out.

Look at the model. It shows how particles probably behave. Each ball represents a tiny particle.

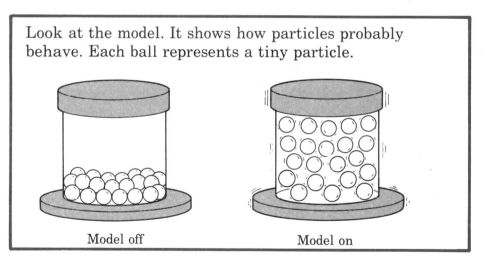

Model off   Model on

1 When the model is switched on, what do the particles do?
2 When the model is switched on, what happens to the distance between the particles?
3 If the particles were given more movement energy what would happen to the spaces between them?
4 Collect and complete the summary diagram. Stick it in your book.

CHECKPOINT

A Fortune teller

When you make a prediction you describe what you think will happen. It is a sort of forecast about the future. A prediction will be more accurate than a guess if it is based on good information and good thinking. A hypothesis can be used to make predictions.

A hypothesis about gravity could lead to this prediction and this guess

---

Use the hypotheses below to make predictions.

Hypothesis 1
We can forecast the future if we know what star sign a person has.

Collect a horoscope.
Read your horoscope.
Predict what is going to happen to you next week **if hypothesis 1 is correct**.

Hypothesis 2
A tossed coin will show heads 50% of the time.

Toss a coin ten times and count the heads. Predict how many times it will be heads in 30 throws **if hypothesis 2 is correct**.

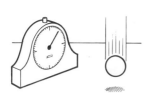

Hypothesis 3
Gravity causes any two balls to fall from the same place at the same speed.

Collect two different balls.
Predict which ball will fall most quickly **if hypothesis 3 is correct.**
Design and do an experiment to find out if the hypothesis is correct. (Think about the fairness of your experiment.)

Describe your opinion of each of the three hypotheses.
(**Hints:** Is the hypothesis a good one? Do you agree with it? Do many people agree with it? Do you think your predictions will be correct?)

B Scientific forecasting

The following drawings represent the spacing of particles in three different substances. Remember that the spaces are big when the particles move a lot. They are small when the particles move a little.

Substance A

Substance B

Substance C

1 Predict which substance will contain particles that move around most.
2 Predict which substance will be easiest to push into a different shape.
3 Predict which substance will contain particles that stay in place on a flat surface.

Collect

Stoppered bottles of substances A, B and C
Empty plastic box

1 Open each of the bottles in turn. Stand the open container on the bench. Decide which substance moves out of the container most easily.
2 Try to pour each substance into an empty box. Decide which substance takes up the shape of the box most easily.
3 Put the box on its side each time. Decide which substance stays in place most easily.

A

B

C

1 Write a short description of each experiment.
2 You made three predictions about substances A, B and C. How many of your predictions were correct?
3 Look at the ten substances in the room. Decide if each substance is like A, B or C. Write these sets into your book.

A Solid, liquid, gas

Particles move. The more they move, the more space there is between them. If the particles are heated then they gain energy and move faster. The particles can therefore change their arrangement. This hypothesis can be shown in a model.

Solid   Liquid   Gas

Your teacher may allow the class to act out the water cycle.

When you hear the word *ice*, you should all act as particles in a **solid**. When you hear the word *water* you should all act as particles in a **liquid**. When you hear the words *water vapour*, you should all act as particles in a **gas**.

Predict which arrangement of particles should be easiest to press together.

Collect

Plastic syringe
Rubber stopper
Wood

1 Fill the syringe with the **gas** air.
2 Hold the end against a rubber stopper and press the plunger hard.
3 Fill the syringe with the **liquid** water. Carefully repeat step 2.
4 Fill the syringe with **solid** wood. Carefully repeat step 2.

air

1 Write a short description of your experiment.
2 Was your prediction correct?
   Explain why particles with this arrangement can be pressed together more easily.
3 Draw your own labelled diagrams to show the arrangement of particles in a solid, a liquid and a gas.

CHECKPOINT

B Hot stuff

The diagram shows a model of the spacing of particles in a solid at different temperatures.

Cold solid

Hot solid

Predict what will happen to
a the length of a solid when it is heated.
b the volume of a solid when it is heated.
c the length and size of a solid when cooled.

Collect
Bunsen
Heatproof mat
Safety glasses
Tongs
Telephone wire
Ball and ring

1 Set up the model telephone wire. Heat it as shown. Find out what happens to the length of the wire.

2 Hold the ball with tongs. Try to put it through the ring. Heat the ball as shown. Find out what happens to the size of the ball.

3 Your teacher will show you a third experiment.

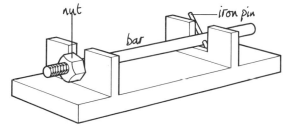

What your teacher will do:
1. Tighten the nut
2. Heat the bar
3. Tighten the nut again
4. Cool the bar

1 Write a short description of each experiment.
2 How many of your predictions were correct?

Build a fire alarm

Heat makes particles in an object move faster. When they move faster they usually get more spaced out. This means that the object expands (gets bigger).

Remember what happens when two metals that are stuck together try to expand by different amounts (see Topic 5.2). Repeat the magic-strip experiment on page 82 if you cannot remember.

Problem
You have to use a bimetallic strip to make an automatic fire alarm. The alarm should ring when there is a heat source nearby.
(Use a bunsen burner as a source of heat.)

Hints
● You will need the equipment shown below.

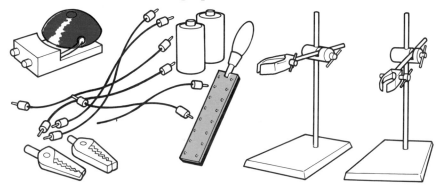

● Your teacher has clue cards to help if you get stuck. Use as few clues as possible.

 Draw a diagram of your final design. Describe how it works.

Good questions

Look at the four pictures. Your teacher will tell one person in your group what has happened in picture A. The rest of the group have to ask this person good questions to get more information about the picture. The person may only answer **yes** or **no** to any question.

The group has a maximum of 20 questions to work out what has happened in picture A. Another person is then told what picture B is, and so on . . .

B What was this object used for?

A A man walks into a bar. He says nothing. The barman takes out a toy gun and points it at him. The man says thank-you and leaves. Why?

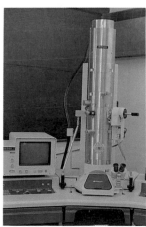

C What is this used for?

D This person always gets off at floor 3, yet she lives on floor 5. Why?

Forbidden ideas

Nicolaus Copernicus

Galileo Galilei

The Earth from the moon

People used to think that the Earth was the centre of the universe. The Earth stayed still and everything moved around it. Two famous scientists had other ideas: **Nicolaus Copernicus** and **Galileo Galilei**.

Nicolaus Copernicus was born in Poland in 1473. He was skilled in many things. He practised medicine amongst the poor. He designed a new money system. He even defended a castle against invaders. Copernicus was also very interested in astronomy. He studied other people's observations of the movement of the planets. These observations did not support the hypothesis that all the planets went round the Earth.

Copernicus thought up, and wrote about, a more likely explanation. The sun was at the centre and all the planets, including the Earth, went round it. His ideas were not liked by the authorities. It was only on his deathbed, in 1543, that he saw printed copies of his book.

His hypothesis did not die. In 1564, a very talented Italian called Galileo was born in Pisa. Galileo was also skilled in many things. He first studied medicine, but then became a professor of mathematics and a successful inventor.

Galileo was fascinated by movement. He proved that falling objects, no matter how heavy or light, take the same time to fall the same distance. In 1609 he improved the recently invented telescope and used it to observe the sun and moon, the planets, and the stars. Many of these observations supported Copernicus's hypothesis.

In 1632 Galileo published a book which discussed the theory of planetary motion. Unfortunately some powerful people did not like his ideas. Galileo was tried and sentenced to house arrest. He died in 1642, a virtual prisoner in his own home.

1 From the passage, describe
   a the hypothesis that Copernicus **disagreed** with
   b the **new** hypothesis that Copernicus wrote about
   c the observations that Galileo made which supported Copernicus's ideas.
2 Use the books in the classroom or from a library to write a paragraph about Galileo, his work and his trial.

6
Science in use

Introducing technology

A Finding solutions

This unit is about science and technology. In technology, scientific ideas and knowledge are used to design, make and test the best solution to people's problems. These can be either design or repair problems (see Unit 6, Book 1). Both types of problem will be solved more easily by understanding science and having a good imagination.

1 What type of problem is it? **2 Any ideas on what to do?** **3 What's the best idea?** **4 How do you try it out?** **5 Did it work?**

Each illustration shows a technological solution to a common everyday problem.
a In each case describe the object in the illustration and the problem that it solves.
b Pick one of these illustrations. Imagine another way of solving the same problem and describe it.

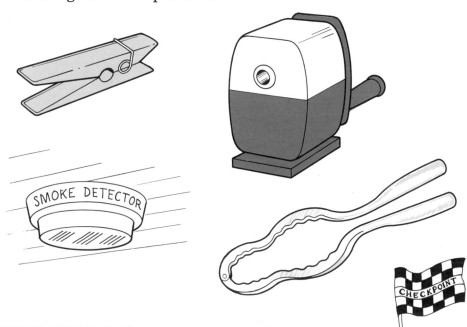

SMOKE DETECTOR

CHECKPOINT

B A race against time

Technology tries to produce the **best** possible solution to a problem.

For example, imagine that you wanted to time a 1000 m race. You could build a sundial using your understanding of science. You could build it in a very scientific way like this . . .

1 Observe a steady change

2 Design a clock

3 Select equipment

4 Build a clock

5 Make one measurement

6 Make a scale

7 Compare with accurate clock

8 Make corrections if necessary

Collect

What you need

Unfortunately a sundial is not the best way of timing this race! It measures time in hours rather than minutes.

Using the same steps as the person above, design and build a better clock for timing the race.

Write a report about your clock.

CHECKPOINT

A Processing

An understanding of science can often lead to ways of solving real problems in the world. For example, scientists have helped to develop the electronic technology which is now used in many devices.

Data handling

Measurement

Detection

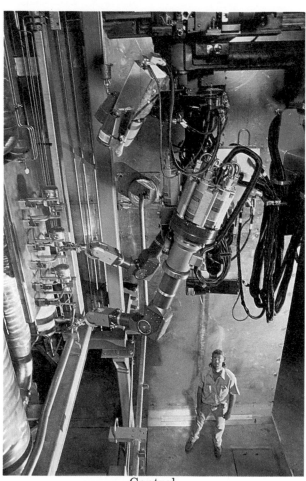

Control

5   Most electronic circuits are complicated. However they make use of some basic ideas.

  1 A current of electricity is made up of particles called **electrons**.

  2 The electrons move (flow) around a circuit.

10 3 The flow of electrons can be
  ● stopped
  ● started
  ● increased
  ● decreased.

Electronic circuits are built up from parts called components. **15**
Some important components are:

 **Switch:** stops and starts current

 **Resistor:** alters the current by a fixed amount

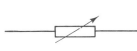 **Variable resistor:** alters the current by a changeable
amount **20**

 **Thermistor:** resistance is changed by heat

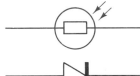 **Photocell or light-dependent resistor (LDR):** resistance is
changed by light

 **Diode:** allows current to flow in one direction only

Transistor: fast acting, sensitive switch **25**

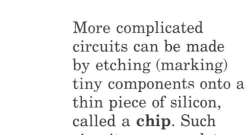

More complicated
circuits can be made
by etching (marking)
tiny components onto a
thin piece of silicon,
called a **chip**. Such **30**
circuits are complete
and yet very small.
They are called
integrated circuits. **35**

Separate microelectronic circuits are used to build up
electronic systems. A system is a collection of circuits that
are designed to work together to do a particular job. The job
that the system is to do is called the **output**. What you need
to put into it is called the **input**. The electronics in the **40**
middle does the **processing**.

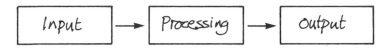

continued ▶

A Processing ▶ *continued*

1 You can **scan** a passage to find certain facts quickly. Each line of the passage on electronics is numbered. Which line or lines mention
   a silicon   **c** electrons
   b output   **d** components?

2 A **table** is a good way of presenting some types of information. Each column in a table has a heading. This heading describes the kind of information in the column. Make a table of three columns to present the information in the electronics passage about common electronic components. The table should show the symbol as well as the job that each component does.

3 An electronic system is described in the passage. It has three parts, called input, processing and output.
In a word processor the input is your typing. The output is what appears on the screen or printer.

What is the input and output of
a a calculator
b a light detector that switches a bulb on when it gets dark
c a siren that goes off when a burglar opens the window?

Your teacher will show you how to use the word processor(s) in the classroom.

1 Type in a summary of the passage about electronics. The summary should have five sentences: one sentence for the main idea in each paragraph.
2 Use the word processor to improve your summary.
3 If possible, print out the final version and stick it into your book.

B Sensors

Electronic circuits are used in sensors. A sensor detects and measures something in the environment. The sensor usually displays the output.

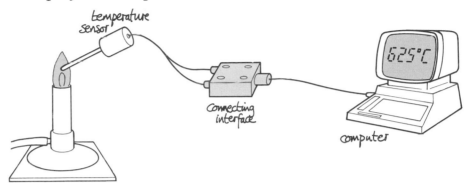

Use one or more of these sensors to measure and display the level of something in your present environment:
- noise sensor
- pH sensor
- light sensor
- resistance sensor
- current sensor
- oxygen sensor
- mass sensor
- time sensor
- position sensor.

Each sensor is set up ready for use, with the instructions for use beside it. Some sensors may be interfaced to a computer.

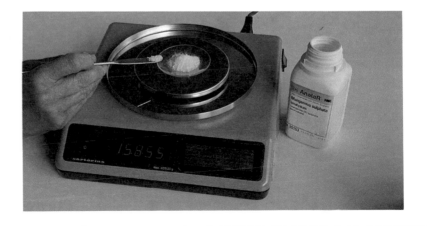

For each sensor that you use, write a short description of what you did and what happened. Be sure to identify
- the input to the sensor
- the type of display that you see.

A Wires and waves

Technology has enabled people to talk to each other and to share information over great distances. Scientists have helped to design and build telecommunication systems.

A telecommunication system has three parts: a transmitter, a carrier and a receiver. The information is sent by the transmitter. It can be carried through wires or optical fibres or even through the air. It is then received and decoded by the receiver. The devices in the cartoon above are mainly receivers.

Transmitter	Carrier	Receiver

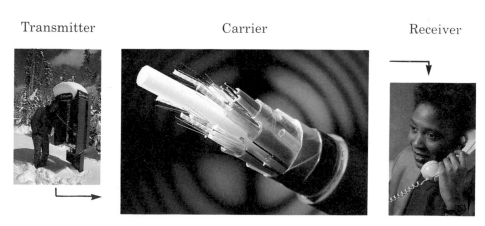

The three parts of an electronic telephone system

Transmitters and receivers are usually complicated electronic systems which contain many integrated circuits. The methods of carrying the information vary.

For example, radio programmes are carried through the air by invisible waves of energy. These waves are passing through you all the time.

Telephone conversations can be changed into electrical currents and carried through wires. They can also be changed into light and carried through very long glass fibres. This means that your spoken words can become bits of electricity or light! In the future even more information is likely to be sent around the world at the speed of light.

1 A picture can hold a lot of information.
   Examine the cartoon of telecommunication devices on the opposite page. List all the devices you can spot.
2 A short passage can only give you a few details, but it often hints at where to look for more.
   a Write down a key word from each paragraph of the passage *Wires and waves*.
   b Use these key words to find out more about telecommunications in the available books. (The index of each book will be helpful.)
      Write down three **new** facts that you discover.

B Teletext

Television sets can now receive more than just sounds and pictures. For example, large data bases are transmitted alongside the programmes of each major TV channel. The BBC system is called Ceefax and the ITV system is called Oracle. Both systems can be received by either a special teletext TV or by a home computer with a special adaptor.

Your teacher will show you how to search a teletext data base. You may even be able to search British Telecom's Prestel data base, which is much bigger.

Use the data base to find the following information. Write each piece of information in your book.

a Today's weather report.
b This evening's BBC1 and ITV programmes (between 18 00 and 21 00 hours only).
c The latest newsflash.
d Two interesting scientific facts.

A Microbes

An understanding of science can help people use technology to solve food problems. Living things are useful in technology. For example, very small **microbes** (which are usually just a single cell) can be used to make food, medicines and fuels. The two most useful types of microbes are moulds (fungi) and bacteria.

Food

People have been using microbes to make some foods for a very long time.

Beer has been brewed for 5000 years. Wild yeast, a single-cell fungus found on the surface of some cereal seed, turns the sugars in the seed into alcohol.

Bread making is another ancient industry. Bakers' yeast is added to the dough. It changes the sugar into alcohol and the gas carbon dioxide. The gas makes the dough rise. When the bread is baked the alcohol is boiled away.

Nearly five hundred years ago the Aztecs of South America made **cakes** from microscopic one-celled plants which they collected from shallow lakes.

As people learnt more about microbes, new technology was invented so that they could be used to produce a wide range of products. Today the **food industry** uses many different microbes.

Bacteria turn milk into yoghurt

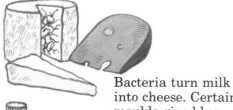

Bacteria turn milk into cheese. Certain moulds give blue cheese its special flavour

Bacteria change beer into malt vinegar

Quorn is a type of protein made from mould. It is used to replace meat in some foods

Enzymes

Enzymes are **special chemicals** found in all living things. Each different enzyme **controls** one type of **chemical reaction.**

The food industry uses enzymes to process many of the foods we eat. It is expensive to get enzymes from plants and animals, but microbes can make large amounts of these chemicals very quickly.

×240

Aspergillus niger

Microbe	Enzyme made	How the enzyme is used
Trichoderma viride	cellulase	to prepare dehydrated vegetables
Brewers' yeast	invertase	to make soft-centred chocolates
Aspergillus niger	lipase	to improve the flavour of ice cream, cheese, chocolate
Aspergillus niger	pectinase	to soften fruit before squeezing the juice out; to make fruit juice clear
Mucor	rennin	to curdle milk in cheese making
Bacillus subtilis	protease	to make meat tender

1 You can **skim** through a passage to get the main ideas quickly. Look for **headings** and words in **bold type** to help you. Usually there is a main idea in each paragraph. Skim the passage on microbes. Write down the three main ideas.

2 You can **scan** a passage to find certain facts quickly. (Look for the same clues as in question 1. Also look at the pictures.)
   a What does *microbe* mean?
   b What are the main groups of microbes used in technology?
   c Write about one very old use of microbes. Include three facts.
   d Write as much as you can about Quorn.

3 A table presents information. You can show the information in another form. For example the flow diagram below shows how the microbe Mucor is used.

Make a similar flow diagram for three of the other microbes in the table. *continued* ▶

▶ *continued*

B Food from respiration

Microbes are used to make some foods. Often the process depends on the **respiration** of the microbe.

For example, during respiration **without** oxygen

- some bacteria use up the sugars in milk and make acid. The milk changes into yoghurt.
- some fungi, like yeast, use up the sugars in grape juice and make alcohol. The grape juice changes into wine. (This process is called fermentation.)

Collect

Thermos flask
Small tin of
evaporated milk
Teaspoonful of live
yoghurt
Plate
Kettle

Making yoghurt

Follow one (or both) of the recipes below.

1 Sterilise flask with boiling water.

2 Add contents of tin plus an equal amount of boiling water.

3 Add one spoonful of live yoghurt.

4 Close flask. Shake. Leave for one day.

5 Pour into dish. Eat.

Collect

Bubble trap
Boiling tube
0.5 g of brewers'
yeast
2 g of glucose
Limewater

Making alcohol

1 Mix the yeast and the glucose.
2 Dissolve the mixture in warm water in the boiling tube.
3 Half-fill the bubble trap with limewater as shown and stopper the tube.

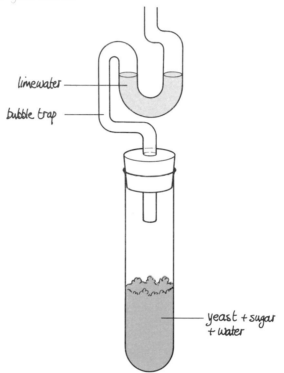

4 Look at the boiling tube after about 5 minutes. Smell the contents after an hour.

1 Draw a flow diagram to show the stages in
   a the yoghurt-making process **and/or**
   b the alcohol-making process
2 In each case write down the observations which show that a chemical reaction has occurred.
3 Look up the meaning of the word *respiration* given elsewhere in this book. (Use the index to help you.) Explain what microbes do during respiration without oxygen.
4 Copy and complete this equation for fermentation.

glucose + yeast → ———— + ————

Chemicals from microbes

A Medical matters

Alexander Fleming

Many living things produce complicated chemical compounds which would be difficult to make in a laboratory. Some of these chemicals are used to help sick people.

One example is the **antibiotic** called **penicillin**. It prevents the growth of many types of bacteria and so helps the human body to fight infection. It was discovered by Alexander Fleming in 1928.

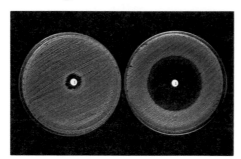

Effect of penicillin on bacterial growth

Penicillin is made from a mould. Huge vats called fermenters are used to make penicillin efficiently. The mould must have perfect growing conditions. The four things that have to be just right are pH, temperature, oxygen level and type of food.

After growing in the fermenter penicillin is separated from the living cells and food by filtration. It is purified by dissolving in a solvent. Next, it is concentrated. The final stage is to add sodium or potassium compounds to allow the formation of crystals.

Another example of a chemical that can be made by a microbe is **insulin**. Insulin is the chemical in our body that controls sugar levels in the blood. A **diabetic** is someone who does not produce enough insulin. **Diabetes** can be treated by taking injections of insulin. Until recently this came from animals. Now it is made by bacteria. The method is shown below.

An example of genetic engineering

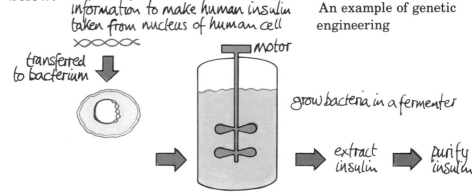

information to make human insulin taken from nucleus of human cell

transferred to bacterium

motor

grow bacteria in a fermenter

extract insulin

purify insulin

1 Skim the passage. Write down the main ideas.
2 A **flow diagram** can be a good way of showing the main steps in a process.
   Copy and complete this flow diagram about penicillin.

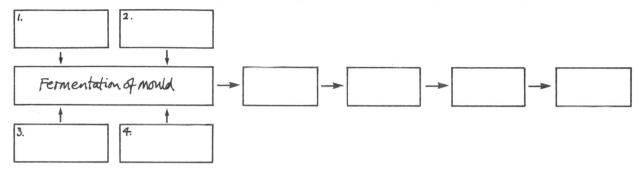

3 Diagrams are often used in science. A good way to test if you understand a diagram is to write a description of what it shows. Describe how insulin is manufactured.

CHECKPOINT

B Build a fermenter

Microbes are often grown in a fermenter. The diagram below shows a batch fermenter for growing brewers' yeast. In this fermenter, the more bubbles that are produced the faster the yeast is growing. (The bubbles are carbon dioxide gas.)

Collect

Fermenter
Bubble trap
Limewater
2 g yeast
Anything else you need

Choice 1
Add spatula-ful of plant nutrient **or** no plant nutrient

Choice 2
Add boiled airless water **or** warm air-full water **or** cold air-full water

Choice 3
Add spoonful of glucose **or** spoonful of sucrose

lime water

2g of yeast

1 Make the three choices shown on the diagram and set up your fermenter.
2 Leave the fermenter for 10 minutes.
   Count how many bubbles are produced in 2 minutes.
3 Record your three choices and your bubble count.

1 Draw and describe your experiment.
2 Compare your results with others in the class. Which three choices help the yeast to grow fastest?

CHECKPOINT

A Types

Materials are substances that are used to make things. An understanding of science is used in technology to design new materials. The drawing shows the main groups of materials.

Metals

Fibres

Plastics

Glass

Ceramics

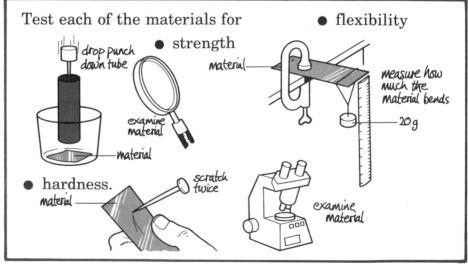

Collect

Examples of each of the groups of materials

Test each of the materials for ● strength ● flexibility ● hardness.

drop punch down tube

examine material

material

material

measure how much the material bends

20 g

scratch twice

material

examine material

Record your results in a table of four columns, headed *Material*, *Strength*, *Flexibility* and *Hardness*.

Some materials like wood and stone are natural: they are found in or on the earth. However, many other materials are synthetic; they are made by people, using technology.

For example, many materials are made from coal and crude oil.

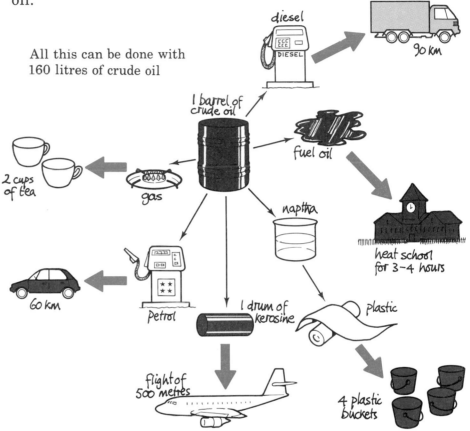

All this can be done with 160 litres of crude oil

The little particles (called **monomers**) in crude oil can be joined together (like beads in a necklace). They become part of a big particle called a **polymer**.

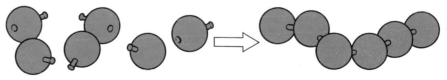

Plastics are all polymers. Plastics are very useful materials.

1 Write down the meanings of the following words: *material, flexibility, polymer, natural* and *synthetic*.
2 What are the main groups of materials?
3 Look around the classroom. List all the useful plastic objects in it.

continued ▶

▶ *continued*

B New materials

This section allows you to make your own materials. Some of these are plastics, some are not. You are going to follow a *New Material* card like the one below. (This is for making nylon and *your teacher* may show you the experiment.)

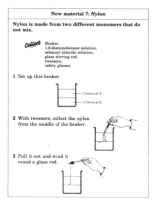

New material 7: *Nylon*

Nylon is made from two different monomers that do not mix.

Collect Beaker,
1,6-diaminohexane solution,
sebacoyl chloride solution,
glass stirring rod,
tweezers,
safety glasses.

1 Set up this beaker.

2 With tweezers, collect the nylon from the middle of the beaker.

3 Pull it out and wind it round a glass rod.

1 **Collect** a *New Material* card.
2 Read the card carefully.
3 Read it again. Make a list of all the equipment and chemicals you will need to make the material.
4 **Collect** the equipment and chemicals on your list.
5 Follow the instructions on the card and make the material.

1 For each material that you make, write out a report like the one below.

Name of material: NYLON
Description of starting substances:
Two poisonous liquids. Clear. One floated on top of the other.
Description of material made:
The nylon was white like wet wool. It looked like a long rope.
Uses:
Used to make clothing, ropes, parachutes, gear wheels, toys.

2 The names of plastics are often complicated but part of the name will often tell you that the plastic is a **polymer**.

Polyamide (nylon)

Polychloroethene (or **Poly**vinylchloride—PVC)
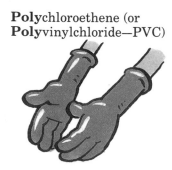

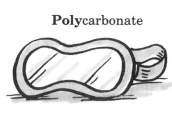
Polyethene

Polycarbonate

Polyurethane

Polystyrene

Polypropene

Polyphenylene

Make a table which shows the names of some common plastics and a use for each.

3 Plastics are used for many different jobs.
The pie chart below shows how plastics are used in Western Europe.

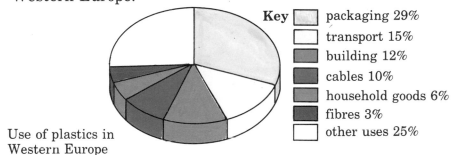

Use of plastics in Western Europe

Key
- packaging 29%
- transport 15%
- building 12%
- cables 10%
- household goods 6%
- fibres 3%
- other uses 25%

Using the figures from the pie chart, copy and complete this bar chart.

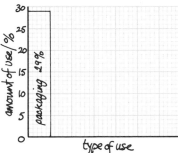

A Old fuels

A fuel is a substance which burns to release energy. An understanding of science has led to better fuel technology. Many natural materials, for example, are refined so that they can be used as fuels.

Sources of some common fuels

Fuel	Source	Use
Petrol	crude oil	
Kerosine	crude oil	
Methane	natural gas (North Sea)	
Charcoal	trees	
Coke	coal	

A good example of a fuel is petrol.

Watch your teacher demonstrate how well a fuel like petrol can burn.

A fuel will only burn when it is hot enough and when oxygen is present.

The fire triangle shows this idea. If any side of it is removed the fire will go out.

Natural fuels will not last forever. The pie chart compares how much of each main type of fuel there is left. The world's oil will probably only last for another 50 years or so.

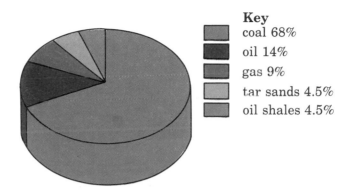

Key
coal 68%
oil 14%
gas 9%
tar sands 4.5%
oil shales 4.5%

Remaining world resources

1 What does the diagram of the fire triangle show?
2 Draw a flow diagram to link the source of a fuel with its use. Use the information in the table opposite. Your diagram will have three boxes.
3 The passage reveals that the world's supply of oil will only last about 50 years. Use the pie chart to estimate how long the world's supply of coal will last.
4 Finding information is an important skill.
   Skim this topic and the previous one to find out what oil is used for.
   Imagine that there is no oil left in the world. Write about the effects this would have.

continued ▶

▶ *continued*

B New fuels

Cars can run on alcohol

Technology has produced different types of fuels. For example, alcohol and methane can be produced in large quantities by microbes. In Brazil petrol is very expensive. Cane sugar is plentiful and is changed into alcohol by yeast. Methane gas is used as a heating fuel. Certain bacteria convert farmyard manure and sewage into this valuable gas.

Methane can be produced from sewage

Collect

Distillation apparatus
Bunsen burner
Liquid from a yeast fermenter

1 The fermenters in the class use growing yeast to convert sugar into alcohol.
   Collect about 20 cm³ of the liquid from a fermenter.
2 Alcohol boils at around 78°C and water boils at 100°C. Use distillation to separate some alcohol from the liquid.

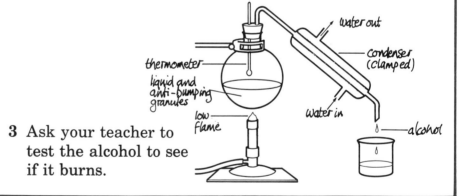

3 Ask your teacher to test the alcohol to see if it burns.

1 Write a report about this experiment. Include **only** the things that you think are important.
2 Making a summary of information is an important skill. In this unit *microbes* have been mentioned several times. Skim the unit. Copy and complete this summary diagram.

Biological washing powder

Biological washing powder contains enzymes which have been produced by microbes. These help to remove protein stains like blood or gravy from clothes by breaking up the big protein particles.

You are going to design an experiment to compare three or four washing powders to see which works best.

Think about how to
  ● use the agar plate (look at page 171)
  ● use the washing powder
  ● make the experiment fair.
Note: The milk agar in the petri dish is white to start with, but if the protein in the milk is broken down the agar will go clear.

1 Write a short report about your investigation.
  Include two diagrams: one to show how you designed your experiment and one to show the result.
2 How could you have made the experiment work more quickly?

Technology

You have to prepare a short talk for the class on one of the following topics. Use the information in this unit and any reference books, newspapers or magazines in the classroom to find **interesting** information. Your talk should last about 2 minutes. (You can practise by recording your talk.)

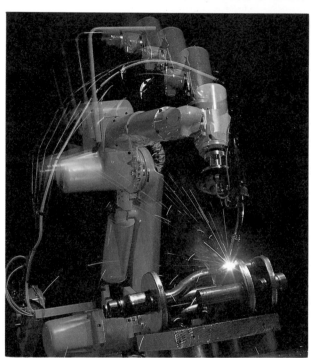

Electronics

Materials

Microbes

Telecommunications

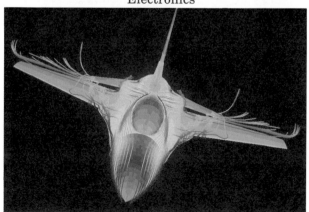

Computers

Fuels

Robot surgeons

Read the following cutting from *The Guardian* (15th August 1989).

Scalpel . . . forceps . . . software . . . as robot surgeon starts operating

Aileen Ballantyne and Peter Large

SURGERY requiring more precision than the human hand can achieve could be performed by robots within the next 10 to 20 years as a result of a Government-funded research project announced yesterday.

The robot would operate the knife and handle probes under the surgeon's instructions for operations such as correcting short-sightedness, the detection and removal of cancerous tumours and prostate gland and joint surgery.

Professor Barrie Jay said it was "not science fiction" to suggest that robots could carry out surgery to correct short-sightedness. "Obviously there would have to be an over-ride for the surgeon, but they could be used for operations which require a very accurate series of cuts within the next 20 years". At present, surgery to correct short-sightedness had "many complications."

Mr Peter Jenkins said the robot's main use would be for movements which could not be achieved by a human hand.

Dr John Dawson was amazed at the thought of a robot-controlled knife or laser operating in the abdomen, where there were wide differences in fatty layers and blood-vessel distribution.

The surgeon's skills in such cases went far beyond the craftsman aspects. Robot help for the surgeon might be useful in more uniform areas like the eye but, even there, basic computer science—let alone artificial intelligence—still could not guarantee reliable software running on reliably designed machines.

1 This is a story about new medical technology.
  What would the new technology operate?
2 What does Professor Barrie Jay think this new technology could be used for?
3 What does Mr Peter Jenkins think this new technology could be used for?
4 Does Dr John Dawson think the idea is a good or a bad one?
5 Do you think the idea is a good or a bad one?
  Give your reasons.

Your future

Scientific knowledge and skills play an important part in many jobs today. Here is a selection.

Find and cut out 10 advertisements for jobs likely to need some scientific knowledge and skills. Use these to make up a newspaper page.

---

1 What kind of job would you like to do in the future?
2 How do you think science will help you in the future, even if you do not use it at work?

Extensions

Fossils

Fossils can help us to find out what the environment was like millions of years ago. Some of the fossils found in Britain suggest that the climate was once much warmer. There are three main types of fossil.

Fossil replacements form when an animal or plant dies and is buried by fine sand or mud. Normally the soft parts decay, leaving behind hard bones or shells. Over many thousands of years these hard parts are replaced by minerals. This is the fossil. The surrounding sand or mud changes into sedimentary rock.

Fly fossilised in amber

Actual remains can become fossils in sedimentary rock or amber. Insects can be trapped in the sticky sap or a tree. They do not decay. Over a long time the sap turns into rock-hard amber.

Impressions are formed by plants or animals in soft sand or mud. These can become fossils. Dinosaur footprints are a good example.

Very few fossils show the complete plant or animal. Scientists have to use their knowledge to make sense of what they see.

Collect

Plasticine
Box of objects

1 Make an impression of part of one object.
2 Ask your partner to guess what it is.

1 What are the three main types of fossil? How are they each formed?
2 Copy and complete the flow diagram below to show how a replacement fossil forms.

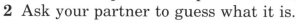

animal dies → ☐ → ☐ → ☐ → fossil

3 Write a short report about the importance of fossils. Use the information here and in other books. Words to look for in an index are *fossils, evolution* and *palaeontology*.

Soil measurements

Look at these four seed packets. The instructions tell you about the types of soil in which the plants grow best.

Plant seeds in open, sunny ground. Likes' well-drained soil.

Plant in shady position. Grows well in chalky soil.

Plant in sunny position. Prefers moist, clay soil.

Plant in open, sunny ground. Must have peaty, well-drained soil.

A soil can be described by

- its appearance (what it looks like)

- its drainage (how much water it holds)   - its acidity (how acid it is).

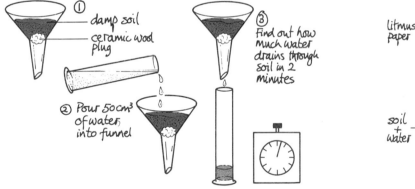

① damp soil — ceramic wool plug

② Pour 50 cm³ of water into funnel

③ Find out how much water drains through soil in 2 minutes

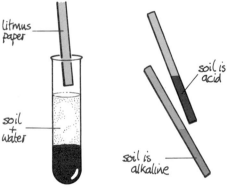

litmus paper

soil + water

soil is acid

soil is alkaline

Collect

Soil chart
Filter funnel
Timer
Measuring cylinder
Boiling tube
Ceramic wool
Litmus papers
Soil sample

1 Use the soil chart to describe the appearance of your soil sample and to find out its type.
2 Follow the experimental instructions to find out about its drainage and acidity.

Write a report about your soil.

Weather forecasting

The layer of air around the Earth is about 200 miles thick. Air is very light but the weight of it pushing downwards can be measured. This is called air pressure. Differences in air pressure cause changes in the weather. When air moves because of pressure differences it is called wind.

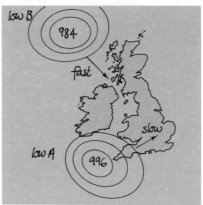

The map shows examples of areas of low pressure (called **lows** or **depressions**) moving over Britain. Some simple rules allow us to forecast the weather for the next day or two.

Rule 1
The weather pattern depends on where the low comes from.

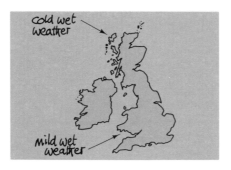

Rule 2
The length of time that the weather lasts depends on how fast the low moves.

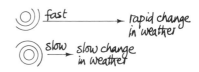

Rule 3
The lower the pressure the more unsettled the weather will be.

1 Look at the map at the top of this page.
Predict the weather caused by the movement of **low A** across the south-east of England.
2 If **low B** now moves towards the south-east of England, what changes in the weather would you expect there?
3 a What is wind?
   b Find out how wind moves around a centre of low pressure. Use books in the classroom or library.

Animal numbers

The animals in an environment are difficult to count. They move around and they often collect in groups. They are not usually spread evenly around the area. They may also be difficult to see, because they are small, timid or camouflaged. The scientist can try to catch a sample of the animals in the environment. The sample is always set free afterwards.

Two useful animal traps are the flowerpot trap and the pitfall trap.

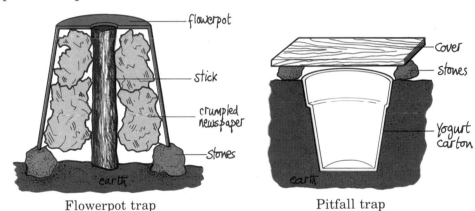

Flowerpot trap   Pitfall trap

Collect

Flowerpot
Newspaper
Stick
Yogurt carton
Trowel
Tray
Hand lens
Animal identity
sheet

1 Build an animal trap in two different places in the school grounds (or in a nearby park). Hide your traps from bigger animals such as dogs (and people!).
2 Leave the traps for one day. Remember where they are!
3 Empty the traps into a large tray. Count the animals.
4 Try to identify some of the animals using the identity sheet (or use a key if one is available).

1 Describe the two places where you set your traps.
2 Make a table of results for each place. Use the headings *Name of animal* and *Number*.
   Give any animal that you cannot identify a letter and draw it underneath the table.
3 Choose one animal from each place. Explain why you think it can survive well in that place.
4 Do you think this way of counting animals in an environment is good or poor? Give your reasons.

Photosynthesis

Photosynthesis is important for many reasons. Green plants make food (sugar and starch) by photosynthesis. They also make oxygen. Animals use up the oxygen. They also eat the food that the plant makes.

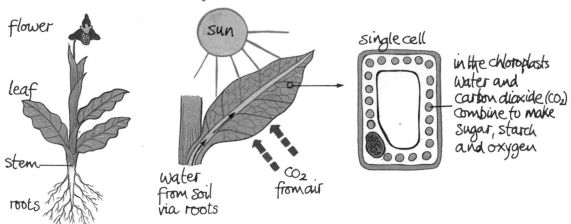

Collect

Water plant
Bench lamp
Metre rule
Timer
Paperclip

Water plants use light energy to make food and oxygen. You can count the oxygen bubbles produced.

1 Examine this drawing of an experiment.

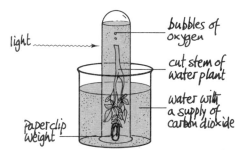

2 Design and carry out an experiment to test whether or not photosynthesis speeds up as the light gets stronger.

1 What are the four main organs of a plant?
2 Write a report about your investigation describing
   ● your experiment and results
   ● what you found out.
3 Copy and complete this sentence.
   In photosynthesis the plant uses _____ energy to change water and _____ gas into _____ and _____ .

Thermal pollution

You may have seen cooling towers at power stations and steel mills. They cool water down before it is returned to a river.

Your task is to find out how warm water could pollute a river by changing the amount of oxygen in it.

Collect

Apparatus shown in the diagram
$50\,cm^3$ of cold water
Bottle of chemical A
Bottle of red dye
$10\,cm^3$ of chemical B
Stirring rod

1 Set up the apparatus as shown in the diagram. Note the scale reading at the start.
2 Slowly add chemical A from the burette. Gently, stir the liquid in the beaker.
3 When the red colour suddenly disappears stop adding chemical A. Note the final burette scale reading. The more of chemical A you add to the liquid in the beaker the more oxygen is present in the water.
4 Repeat the experiment using warm water instead of cold water.

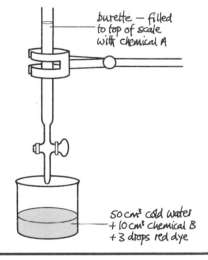

burette — filled to top of scale with chemical A

$50\,cm^3$ cold water + $10\,cm^3$ chemical B + 3 drops red dye

1 What happens to the amount of oxygen in water as its temperature rises?
2 Insects like mayfly larvae use gills to breathe in water. Why will their gills beat faster in warm water than in cold water?
3 Why do you think this experiment is called *Thermal pollution*?

Breathing and burning

Oxygen is a gas. It is found in the air. Oxygen helps your body get energy from food (**respiration**). Oxygen is also needed to get things to burn and release energy (**combustion**).

Respiration

Combustion

Collect

Test tube of oxygen
Wooden splint
Bunsen burner and mat
Equipment for each experiment
Safety glasses

1 **The test for oxygen**
  a Light a wooden splint.
  b Blow it out.
  c Put the **glowing** splint just inside the test tube.
    Observe what happens to the end of the splint.
    This is the test for oxygen.

2 Use the test to find out which of these experiments produces oxygen.

1 How do you test for oxygen?
  What happens?
2 Write a short report about the experiments you did.

Useful properties of metals

Metals are useful because they can be
- made into wires
- bent and shaped

Some metals make stronger wires than others. Some metals change shape more easily than others.

1 Study the diagram. It suggests a way of comparing the strength of metal wires.

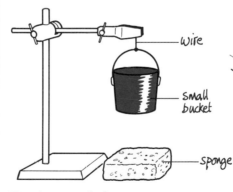

Design and do an experiment which compares the strength of the wires.
Make sure that your comparison is **fair**.

2 Study the diagram. It suggests a way of changing the shape of metal strips.

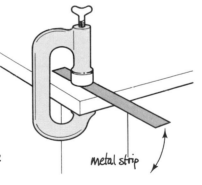

Use this method to compare how well the strips keep their shape. Make sure that your comparison is **fair**.

For each experiment
- draw your equipment
- present your results in a table with two columns
- describe how you made the comparison fair.

Metals and non-metals

Metals are used throughout a modern home. See how many examples you can spot in this cartoon.

Metals have useful properties.
Metal elements are on the left-hand side of the periodic table. They are all solids at room temperature, except mercury. The 21 non-metallic elements are on the right-hand side. Only nine are solids at room temperature. Metals have other important properties that can be found by experiment.

Collect

Tray containing solid elements (copper, iron aluminium, zinc, sulphur, carbon, iodine)
Anything else you need to do the tests

1 Divide the elements into two sets by appearance **only**.
2 Carry out experiments to find out
  a which elements conduct electricity
  b which elements conduct heat from hot water easily
  c which elements break into smaller pieces easily.
  In each case write down your results.

1 Look at the periodic table. List the elements that you tested under two headings: metals and non-metals.
2 Describe these properties of the metal group: appearance, conduction of electricity and heat, and strength.
3 Describe the same properties of the non-metallic group. (*Be careful:* there is at least one unexpected result in this group.)
4 Pick one property of each group.
  What could an element with this property be used for?

Electroplating

Electricity can be used to separate an element from a compound. In this way, a metal element can be attracted to an object so that it forms a metallic coat. This is known as **electroplating**.

These objects have all been electroplated

How to electroplate a coin

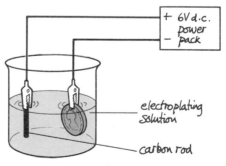

1 Set up the circuit.
   Use copper sulphate solution in the beaker if you have a 'silver' coin.
   Use nickel sulphate solution in the beaker if you have a 'copper' coin.
2 Make sure that the coin is the negative terminal.
3 Switch on and leave for a few minutes.

Describe the experiment.
Include a labelled diagram.

Using up oxygen

Burning and rusting both use up oxygen and a new compound called an oxide is formed. They use oxygen which is found in the air. This fact can help us to work out how much of the air is made up of oxygen.

Collect

Two boiling tubes
Taper
Iron filings
Beaker
Bunsen burner and mat
Plasticine

Burning

1 Set up this apparatus.

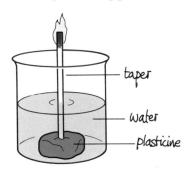

taper
water
plasticine

2 Light the taper. **Quickly** put the boiling tube over the flame. Observe.

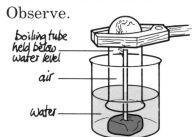

boiling tube held below water level
air
water

Rusting

1 Set up this apparatus.

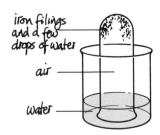

iron filings and a few drops of water
air
water

2 Leave the experiment for about one week. Observe.

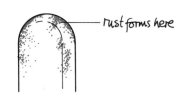

rust forms here

1 For **each** experiment
- Draw the result. (Show the final level of water in the tube as accurately as you can.)
- In each case the oxygen in the boiling tube was used up. Estimate what fraction of the air in the boiling tube was used up (e.g. $\frac{1}{6}$, $\frac{1}{4}$ and so on).

2 Oxygen makes up about one-fifth of the air.
   a Which experiment used up one-fifth of the air in the tube?
   b Give a reason why the other experiment did not use up one-fifth.

Plant indicator

Some plants contain chemicals that change colour in different soils. The hydrangea has blue petals in acid soil and pink petals in less acid soil.

Collect

Plant material
Bunsen burner and mat
Tripod stand and gauze mat
Beaker
Dropper
Set of solutions

1 Boil the plant material with about 40 cm³ water in a beaker for a few minutes.

2 Add several drops of the coloured liquid to each of the solutions shown in the test tubes below.

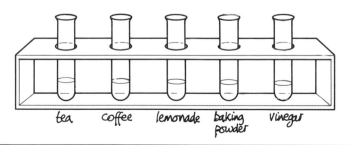

1 Describe how you made the indicator.
2 What colour is your indicator in acid and in alkali?
3 Make a table to show your results.

If you have time try to make another plant indicator.

A chemical circle

Limestone is a sedimentary rock, mostly made from the shells of tiny sea creatures that died a very long time ago. Limestone contains the compound called calcium carbonate which can be changed by several chemical reactions.

Collect

Limestone
Straw
Dropper
Filter funnel and paper
Anything else you need
Safety glasses

Chemical reaction 1

1 Heat the limestone piece very strongly on a wire gauze for several minutes. If you heat it strongly enough it will glow.

Chemical reaction 2

2 Leave the rock to cool. This rock is called lime.
3 Put it in a glass beaker. Add several drops of water. Listen carefully. Touch the bottom of the beaker.

Chemical reaction 3

4 Add about 20 cm³ of water to the lime. Filter the solution into a flask. The solution is called limewater.

5 Bubble your breath into the solution through a straw. Watch carefully. The carbon dioxide in your breath reacts with the limewater. The new substance that forms is calcium carbonate.

1 For each reaction explain how you knew that a chemical reaction was happening.
2 Copy and complete the diagram below. Use it to explain why this is a chemical circle.

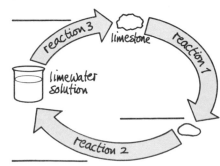

All systems go

The human body is made up of several different organ systems. These systems do particular jobs but in a healthy body all of the systems work together. Here is a problem that the body must solve using many different systems.

How does the body get energy from food?

The diagram summarises how this problem is solved. It shows what happens and which main organ systems are involved.

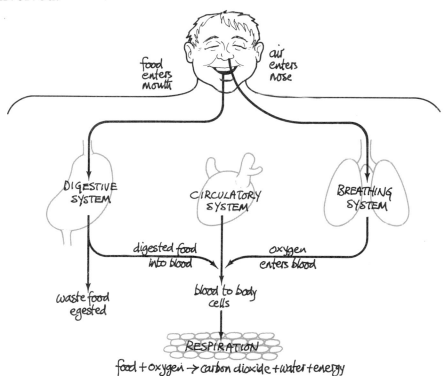

1 What are the three main body systems involved in the release of energy from food?
2 Where in the body is energy released?
3 What is the name of the chemical process used for energy release? Describe this process.
4 What is the energy used for?
5 Use the diagram to explain why we breathe out more carbon dioxide and water vapour than we breathe in.

Weight watcher

You are young, growing and active so you need lots of energy and lots of protein. The amount of energy you need depends on many things like your age, size, how active you are, and if you are a boy or a girl.

However, your weight gives a rough guide to the amount of energy and the amount of protein that you need each day. You need

- about 170 kilojoules of energy every day for each kilogram of body weight
- You need about 0.6 g of protein every day for each kilogram of body weight.

Collect

Weighing scales
Food wrappers with nutritional information

1 Weigh yourself.
2 Work out your energy needs in kilojoules per day.
3 Work out your protein needs in grams per day.
4 Design a meal using the nutritional information on the food wrappers.
5 Calculate the total energy supplied and the total protein supplied by this meal.

1 What affects the amount of energy you need each day?
2 Why do you need a lot of protein?
3 Copy and complete this table.

My weight (kg)	Energy per day (kJ)	Protein per day (g)

4 List the foods you chose and the energy and protein supplied by each one. Show the totals.
5 What would happen to your weight if your daily energy intake was
  a much less than your energy needs
  b much more than your energy needs?

Fluoride and tooth decay

Substances containing fluoride are found in some natural rocks. It sometimes gets dissolved into the drinking water. A survey in the USA found a link between the amount of fluoride in children's drinking water and the number of decayed teeth that the children had.

The graph shows the results of the survey. More than 7000 children took part.

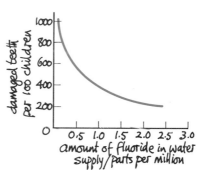

Similar surveys were done in the United Kingdom. They showed that water with more than one part of fluoride per million reduced the number of damaged teeth by about 60%. Therefore, some water authorities now add fluoride to drinking water and many toothpastes have fluoride added to them.

However, there are people who argue against adding fluoride to drinking water. They say that fluoride is harmful. A dose of 2500 parts per million will kill an average person. They also say that fluoride does not work. Finally they say that adding substances to water is interfering with nature, and with their freedom to choose.

1 Why do some people want to add fluoride to drinking water?
2 Why do some people argue that fluoride should not be added to drinking water?
3 From the graph, how many teeth are damaged in 100 children when the amount of fluoride is
   a 0.5 parts per million
   b 1.0 parts per million?
4 Discuss the argument with your partner.
   What do you think? Should fluoride be added to **your** water? Find out whether it is.

Digesting fat

All food substances are broken down and made soluble by special biological chemicals called **enzymes**. Enzymes are substances that help chemical reactions to take place in living tissues. **Lipase** is an example of a fat-digesting enzyme. The word equation below shows that two new substances are formed.

$$\text{fats} \longrightarrow \text{fatty acids} + \text{glycerol}$$

If lipase is added to milk, which is rich in fats, you can't see anything happening. If you add lipase to a mixture of milk and universal indicator you can see something happening.

Collect

Bottle of milk
Bottle of lipase
Bottle of universal indicator
Beaker of warm water
Two 5 cm³ plastic syringes
Two test tubes
Marker pen
pH colour chart

1 Put 5 cm³ of milk into two test tubes labelled 1 and 2, using a syringe.
2 Put 10 drops of indicator into both test tubes. Shake the tubes.

3 Place both test tubes in the beaker of warm water.
4 Add 2 cm³ of lipase to test tube 1 with the other syringe.
5 Look for a change after 15 minutes.

1 Write a report about your experiment.
Explain what has happened. The information at the top of this page should help you.
2 If you have time, design and carry out an experiment to find out if temperature affects the speed of this reaction.

Drug abuse

When chemical substances that are useful to us are misused, they can become dangerous. When drugs are taken or solvents like glue or gas are sniffed, chemicals are passed by the blood system to the brain. At first the person may feel good, but this feeling soon wears off. Drugs can make people lose control of their actions. Any damage caused by drugs is difficult to cure.

Some results of a survey of over 1000 fifteen-to-sixteen year olds are shown in the table below. The dangerous side effects of the drugs are also shown.

Drug	% Male abusers	% Female abusers	Effects
Cannabis	7.4	7.1	Similar to alcohol. Reactions slowed down, co-ordination poor.
Tranquillisers	5.4	5.1	Drowsiness, slow reactions. Addiction is possible.
Glues/solvents	5.4	4.0	Danger of addiction. Can cause heart damage, suffocation and death.
Other drugs	4.3	3.8	Many other drugs such as heroin and cocaine are addictive. They can cause death.

The survey also found that people who tried drugs were likely to be people who already smoked and drank alcohol.

1 Draw a bar chart to show the information in the first three columns of the table above. Make a colour key for your bar chart.
2 Write a letter to a friend to try and persuade them to avoid taking one of the substances from the table.

Your activity clock

Fit people are usually active people. They may play sports. Sport can develop **stamina**, which means that you have staying power. Sport can also develop **suppleness**, which means that you can bend and twist easily. Finally it can develop **strength**, which means that your muscles are powerful. Here is a list of activities with a rating for each one.

	Stamina	Suppleness	Strength
Badminton/tennis	✓✓	✓✓✓	✓✓
Canoeing/rowing	✓✓✓	✓✓	✓✓✓
Cycling (hard)	✓✓✓✓	✓✓	✓✓✓
Dancing (disco)	✓✓✓	✓✓✓✓	✓
Football/hockey	✓✓✓	✓✓✓	✓✓✓
Golf	✓	✓✓	✓
Gymnastics	✓✓	✓✓✓✓	✓✓✓
Hill walking	✓✓✓	✓	✓✓
Jogging	✓✓✓✓	✓✓	✓✓
Judo	✓✓	✓✓✓✓	✓✓
Sailing	✓	✓✓	✓✓
Squash	✓✓✓	✓✓✓	✓✓
Swimming (hard)	✓✓✓✓	✓✓✓✓	✓✓✓✓
Weightlifting	✓	✓	✓✓✓✓

Key

✓	no real effect
✓✓	beneficial effect
✓✓✓	very good effect
✓✓✓✓	excellent effect

You can get some idea of your fitness by charting your level of daily activity. The picture shows activity clocks for a fit person and an unfit person.

SLEEPING
RELAXING
LIGHT EXERCISE
ACTIVE EXERCISE
EATING SNACKS
EATING MEALS

1 What three things does sport help to develop?
2 What is the difference between the activity clocks of a fit and an unfit person?
3 **Collect** a blank activity clock.
   Complete the clock for last Saturday.
   Stick it into your book.
4 Which activities do you enjoy?
5 What could you do to improve your fitness?

Alcohol survey

Not many people fully understand the dangers of alcohol. This may be because they don't know about how alcohol affects their body. Perhaps they don't know how much alcohol they are actually drinking.

Your task is to design a survey questionnaire to find out what people in the school really think about drinking alcohol.

Here are some suggestions to get you started.

- Be sure that you know the facts. Look back at page 57.
- Write questions that can be answered **yes** or **no**.
- Find out what people know. For example,
  Does a half pint of beer have the same amount of alcohol as a measure of whisky? (yes)
  or
  Can your body get rid of the alcohol in a pint of beer in less than 1 hour? (no)
- Find out what people think about drinking alcohol. For example,
  Do you think that the legal limit for driving should be lowered from 80 mg of alcohol per 100 ml of blood?

Collect

Clipboard
Paper
Poster paper
Coloured pencils

You and a partner have to think of ten questions to ask. Discuss them first.
Decide who to ask.

1 Write down the ten questions.
2 Conduct the survey.
3 Record your results on a form like this.

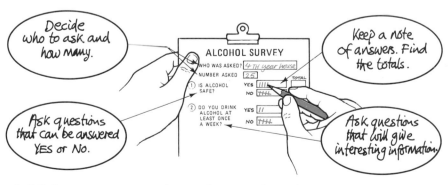

4 Make a poster to show the results of your survey.

Reaction times

When you see something happen, you react to it. For example, if you are riding your bike and someone runs in front of you then you put the brakes on. This reaction takes about 0.30 s. This is the time it takes for a message to go from the eye to the brain, for the brain to decide what to do and for another message to come back from the brain to the hand muscles.

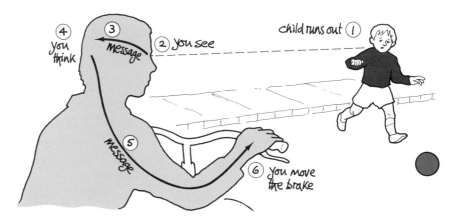

You can learn to speed up your reaction time. You can also slow down your reaction time by confusing your brain with other things.

1 Measure and record your reaction time four times in a row using **one** of these methods
   - a reaction timer machine
   - a computer program
   - a ruler (see the information sheet).
2 Now measure and record your reaction time as you count backwards from 23 to 13.

1 What is *reaction time*?
2 Describe how you measured reaction time.
3 Draw a bar chart which shows the reaction times for your first try, second try and so on.
   a Did you improve?
   b Why is it useful to practise a sport like tennis or badminton even when you are fully fit?
4 **a** Did counting affect your reaction time?
   b Why should drivers concentrate when they are driving?

Measuring forces

Length is measured in units called centimetres. Temperature is measured in units called degrees Celsius. Pushing, pulling, tearing and twisting forces are measured in units called **newtons**. The name comes from a very famous scientist, **Sir Isaac Newton.**

Isaac Newton was born in 1642. He was a brilliant mathematician and scientist. He studied forces and movement. His laws on motion are still used today.

Some forces can be measured using a **newton meter**.

Isaac Newton

Collect

Newton meter
Trolley
Masses

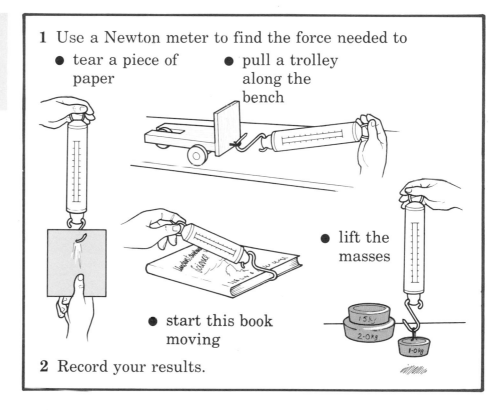

1 Use a Newton meter to find the force needed to
- tear a piece of paper
- pull a trolley along the bench
- start this book moving
- lift the masses

2 Record your results.

1 What unit of force could you measure your weight in?
2 What instrument is used for measuring force?
3 Plot your results for lifting 1 kg, 1.5 kg and 2 kg masses as a line graph. Label the axes *Mass* and *Force*.
4 Use your graph to work out the force needed to lift
   a 1.75 kg, **b** 2.5 kg.
5 Predict what force would be required to lift
   a 17 kg, **b** 21.7 kg, **c** your own mass.

Electromagnets

Many household gadgets contain electromagnets.

The drawing in the box below shows a simple electromagnet. Some wire has been wound around a piece of iron. When an electric current flows through the wire coil in one direction the iron becomes a strong magnet. When the current is switched off again the iron loses its magnetic force.

Collect

Length of wire
3 cells
1 connecting wire
Iron nail
Switch
Box of small pins

1 Make an electromagnet like the one shown in the picture. Use ten coils of wire and **one** cell. Find out how many pins it can pick up.
2 Try to increase the strength of your electromagnet. (There are some hints in the picture.)
Test the strength of the electromagnet by counting the number of iron pins that it can pick up.
Try out as many ideas as you have time for.

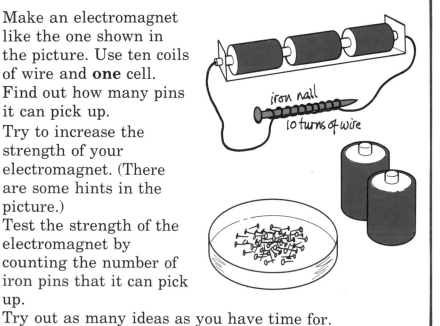

iron nail
10 turns of wire

1 What is an electromagnet made from?
2 Give three examples of gadgets that use electromagnets.
3 How can you turn an electromagnet on and off?
4 Describe how you improved your electromagnet.

Stop!

A force is needed to stop a moving object. In a car the brakes use friction to slow it down. The faster a car is moving the longer it takes to stop, and so the stopping distance is greater. This is important in road safety.

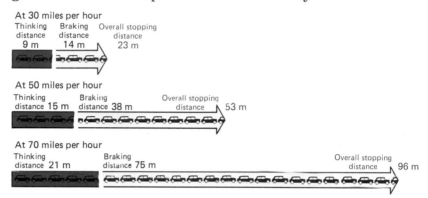

At 30 miles per hour
Thinking distance 9 m   Braking distance 14 m   Overall stopping distance 23 m

At 50 miles per hour
Thinking distance 15 m   Braking distance 38 m   Overall stopping distance 53 m

At 70 miles per hour
Thinking distance 21 m   Braking distance 75 m   Overall stopping distance 96 m

Stopping distances for a car travelling at various speeds on a dry level road (information from *The Highway Code*)

Collect

Sheet of graph paper

Draw a line graph to show how stopping distance changes with the speed of the car.
Remember to put scales on the graph and to label both axes. Give your graph a title.

1 In wet weather the stopping distance of a car is much greater. Why is this?
2 The picture shows an emergency stopping lane on a hill. Explain how it can stop a car whose brakes have failed.

3 Car brakes get very hot in use. Where does this heat energy come from?

Levers

Imagine trying to build an Egyptian pyramid in the school playground. You are not strong enough to produce the force necessary to lift a single stone. This does not mean that you could not do it. All you need is a **lever** and a **pivot**.

A lever seems to allow us to increase the force we can produce. This idea of multiplying force is used in many simple machines.

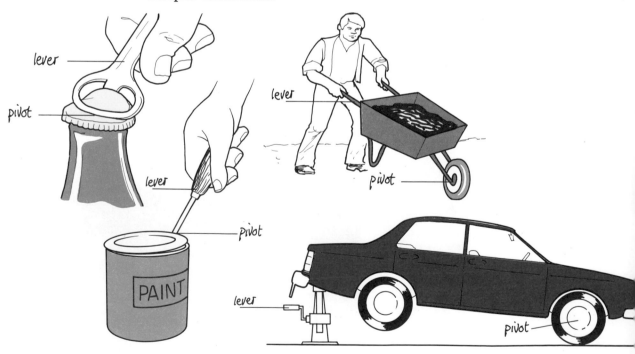

lever

pivot

lever

lever

pivot

lever

pivot

lever

pivot

continued ▶

You can use a see-saw to learn more about how a lever works.

Collect

See-saw (made from a lever and a pivot)
7 small masses

1 Set up your see-saw.
Experiment with the lever and pivot to solve these balancing problems. There is usually more than one solution. You can place the masses wherever you like. You can place the pivot wherever you like.

2 Try to balance the masses in piles as follows:
   ● **4** on the left of the pivot with **3** on the right
   ● **5** on the left of the pivot with **2** on the right
   ● **6** on the left of the pivot with **1** on the right.

1 Give three examples of a lever in action.
2 Copy and complete the table below to show the results of your experiments

Left side		Right side	
Number of masses	Distance from pivot (in squares)	Number of masses	Distance from pivot (in squares)
4		3	
5		2	
6		1	

3 What is the pattern in your results?
   (*Hint:* Multiply the mass times the distance on each side of the pivot. Now you should be able to answer the question.)
4 Where would you place a 2 g mass to balance the see-saw if the other side had:
   a a 10 g mass 2 cm from the pivot
   b a 6 g mass 6 cm from the pivot?

Moving designs

When bones at a joint move, two sets of muscles act against each other. This is true in all animals with joints. In this picture the muscle that lifts the lower leg is shown in green and the muscle that straightens the leg is shown in red.

Flick a tail
Fish swim by flicking their tails from side to side.

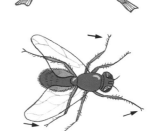

Shake a leg
Insect legs have joints which move when they walk. However, their skeleton is on the outside and their muscles are on the inside!

Flap a wing
Birds fly by flapping their wings. The diagram on the right shows the skeleton from the front. Both muscles are attached at one end to the large breastbone.

Collect

Movement worksheet
Red and green crayons

1 Complete the drawing on the worksheet of the fish to show the muscle used to flick the tail to the right (in red), and the muscle used to move the tail to the left (in green). (The ×s give you a clue about where the muscles join the bone.)
2 Complete the drawing on the worksheet of the insect leg to show the muscle used to bend the leg (in red), and the muscle used to straighten the leg (in green).
3 Complete the drawing on the worksheet of the bird skeleton. Draw the muscle that would lower the wing in red. Draw the muscle that would raise the wing in green.
4 Label all the muscles you have marked. Stick the worksheet into your book.

Spoilt for choice

It is very important to choose the best materials for the job. To do this you have to match the properties of the finished object with the properties of the materials.

If you want a car body to be strong you might decide to use steel. If the car body has to be light you might decide to use a strong plastic. Today there is a wide range of plastics. Some of these are shown in the table below.

Plastic	Property					
	Spongy	Tears easily	Floats in water	Scratch with fingernail	Melts easily	Catches fire
PVC			✓	✓	✓	✓
Melamine						✓
Expanded polystyrene	✓	✓	✓	✓	✓	✓
Polypropylene		✓			✓	✓
Polyethene					✓	✓

Here are four plastic products. Discuss what important property each **must** have.

 a

 b

 c

 d

Copy and complete the table below.

Plastic product	Choice of plastic	Reason
a.		
b.		
c.		
d.		

Clear thinking

Problems are solved by thinking and imagination. If you can think clearly then you will be more likely to solve a problem. Clear thinking often takes time. Take your time during this experiment.

Collect

Stand
2 magnets
Cling film

Experiment 1
Slide both magnets down the rod with the label upwards.

Experiment 2
Turn one of the ring magnets so that the label is facing downwards.

Experiment 3
Cover the magnets with cling film and repeat the two experiments.

These five pupils also did the three experiments. Here are their results and comments.

	Carmel	Barry	Albert	Delia	Ela
Experiment 1	The two magnets went to the bottom.	One magnet got stuck.	The top magnet didn't touch the bottom one.	There was a force between the magnets.	The two north poles forced each other apart.
Experiment 2	The two magnets went to the bottom.	The magnets had their south poles touching.	The two magnets stuck together.	There was a force between the magnets.	The sides with the labels go together
Experiment 3	Magnetism gets through cling film.	The magnetic force is reversed when the magnet is upside down.	Gravity pulls the magnet down.	There is a force between the magnets which acts at a distance.	Cling film stops magnetism

1 Which of the five pupils got the same results for experiment 1 as you did?
2 Which of the five pupils describes something you did **not** see in experiment 2?
3 The five pupils tried to explain their results for experiment 3. Which three explanations could be correct?

Coloured compounds

It is not easy to find a pattern in things which are new to you. If there are a lot of objects or observations then the pattern is even more difficult to find. It is then useful to look at two or three bits of information first.

For example, some coloured compounds are shown below. The name of each compound is made up of two words; so *sodium chloride* is made up of *sodium* and *chloride*.

1 Sodium chloride

2 Copper chloride

3 Sodium chromate

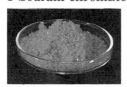
8 Nickel sulphate

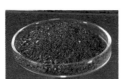
4 Potassium permanganate

5 Copper sulphate

6 Potassium chloride

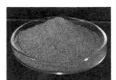
7 Ammonium dichromate

9 Potassium chromate

10 Nickel chloride

11 Ammonium sulphate

12 Cobalt sulphate

1 What colour do you think **sodium** is linked with? (Look at compound 1.)
2 What colour do you think **chloride** is linked with? (Look at compounds 1 and 6.)
3 What colour do you think **copper** is linked with? (Look at compounds 2 and 5.)
4 Find other patterns in the colours of these compounds and report the patterns in a table.

Word in the name	Colour

Work out what colour the following compounds will be. Write your ideas down. Then go and check your answers by looking at the compounds.
- Cobalt chloride
- Ammonium chloride
- Sodium dichromate

157

Asking questions

In many situations it is easier to work out the likely explanation if you know how to ask good questions. Some problems can be solved by asking questions which seek **important** details.

Look at the drawings below. Each one shows a very unusual situation. Your teacher knows what has happened in each case. The class have to try to work out what has happened. Put your hand up to ask questions. Your teacher can only answer **yes** or **no**. The class that asks good questions will solve the puzzles.

How was the elephant hurt?

How did I change the light bulb?

Cleopatra is asleep in bed. Mark Anthony is lying dead on the floor. There is some broken glass and water near him. How did he die?

Atoms and molecules

You studied elements and compounds in Unit 2. **Elements** are substances that are made up of one kind of particle, called an **atom**. **Compounds** are made up of two or more kinds of atoms joined together in a **molecule**.

 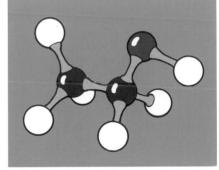

The **element** carbon   The **compound** ethanol

Collect

Model-making kit

1 Make a model of a
   water molecule.

The symbol for water is H_2O because there are two hydrogen atoms (white) joined to one oxygen atom (red).

2 Make models of the following substances.

Substance	Formula
Hydrogen	H_2
Methane (North Sea gas)	CH_4
Ammonia	NH_3
Carbon dioxide	CO_2

1 What is an atom?
2 What is a molecule?
3 Draw two of the molecules that you built.
   Write the name and symbol beside each one.
4 Imagine that you are a water molecule inside a kettle.
   Describe what happens to you when the kettle is switched on and left to boil dry.

Ammonia on the move

Ammonia can be detected by using pH paper. The paper changes colour when ammonia particles touch it.

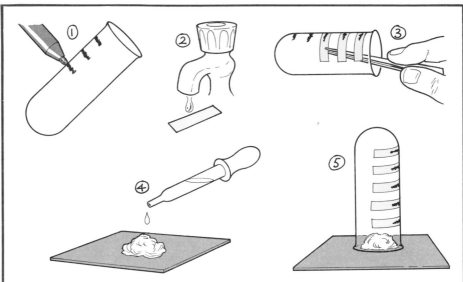

1 Mark the outside of the boiling tube every 2 cm, beginning at the open end.
2 Wet the pieces of pH paper.
3 Put the wet paper inside the boiling tube at each mark. Use the tweezers.
4 Wet the cotton wool with ammonia. Put it on the heatproof mat.
5 Cover the cotton wool with the boiling tube. Time how long each piece of pH paper takes to change colour.

1 What colour does wet pH paper change to when ammonia touches it?
2 Copy and complete the results table.

Distance from open end of tube (cm)	Time taken to change colour
2	
4	
6	
8	
10	

3 What must the ammonia particles be able to do to cause **all** the pieces of paper to change colour?

Diffusion

Diffusion is the word which describes how particles of one substance move through another substance.

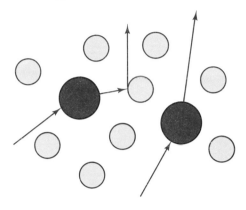

Collect

1 crystal of iodine
1 crystal of potassium permanganate
2 test tubes
Test tube rack

1 Fill the test tubes with water.
   Put them in the test tube rack.
2 Drop the iodine crystal into one tube and the potassium permanganate crystal into the other.
3 Watch what happens. Do not disturb the test tubes for several minutes.

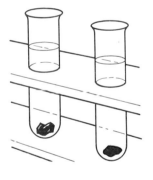

1 Write a report about your experiment. Include a diagram of the results.
2 Which substance is best at diffusing through the water?
3 Write down any reasons which could explain why some particles can diffuse through water faster than others.
4 If there is time, your teacher may show you how fast a gas called bromine can diffuse. (**Note:** Bromine is **dangerous**.) Describe this experiment.

The solar system

The solar system is the name given to the sun and all the planets which go round it. The drawing shows the arrangement of planets in the solar system.

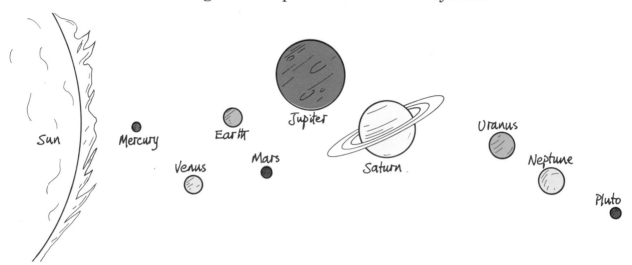

Astronomers study the stars and planets. They have worked out the size and densities of all the planets.

Planet	Size (relative to earth)	Density (kg m^{-3})
Mercury	0.4	5400
Venus	0.9	5300
Earth	1.0	5500
Mars	0.5	3900
Jupiter	11.2	1300
Saturn	9.5	700
Uranus	4.0	1300
Neptune	3.9	1600
Pluto	0.2	2100

1 Draw a bar graph to show the sizes of the planets. Plot the planets in the same order as the table.
2 Draw a similar bar graph to show the densities of the planets.
3 Compare your two bar graphs and look for patterns.
  a How many groups can you divide the planets into?
  b Write down a hypothesis which describes a connection between the size and density of a planet.
4 Which planet does not fit your hypothesis?

Solid, liquid, gas

Substances come in all shapes and colours. However, they all belong to one of the three sets of substances called **solids**, **liquids** and **gases**.

1 There are lots of substances laid out around the room. Each of them has a number. Go to each one in turn. Write down the number of the substance and its name (if you know it). Also write down which set of substances you think it belongs to.
2 When you have finished, join a discussion group of three or four people.
The group has to discuss the answers to three questions.
- What is similar about all solids?
- What is similar about all liquids?
- What is similar about all gases?

Write your ideas about the discussion questions under these headings.
- I know a substance is a solid because . . .
- I know a substance is a liquid because . . .
- I know a substance is a gas because . . .

Expansion

A solid usually gets bigger when it is heated. This is called **expansion**. The opposite of expansion is **contraction**. Solids expand because the particles get more energetic when they are heated. They move more and so the spaces between them get bigger. The particles do not get bigger.

Collect

Conical flask
Long glass tube and stopper
Marker pencil

1 Fill the flask with water. Stopper it as shown. Make sure that the water level just shows above the stopper. Mark this level.

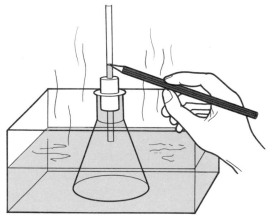

2 Place the flask in hot water. Leave it for 2 minutes.
3 Mark the new water level.

1 Use a dictionary to find and write down the meanings of the words *expand* and *contract*.
2 Write a report about this experiment. Include a description of what happened to the water level in the flask when the water was heated.
3 Write a likely explanation for your observations.
4 Describe an experiment that you could do to find out if liquids contract when they are cooled. Try your experiment.

Design solutions

Design (but **do not** build) a solution to **two** of the following problems. Write about your design. Include a labelled sketch of it.

1 An electronic calculator that a blind person can use.

2 A way of watering your indoor plants when you are away on holiday.

3 A way of stopping a toddler from opening a cupboard **without** fitting a lock.

4 A device for dividing this cake into **six equal** pieces.

Truth tables

There are two ways of sending information through electrical circuits.

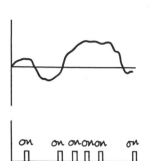

In **analogue circuits** there is a range of information. Analogue information is used in radio transmission, LP records, TV transmission and telephone messages through wires.

In **digital circuits** there are only two bits of information.
● **off** is given the symbol **0**   ● **on** is given the symbol **1**

Digital information is used in laser discs, compact discs, telephone messages through optical fibres and equipment that takes decisions (like computers).

A digital circuit can be set up using two switches.

1   Build a circuit which has a bulb, a battery and two simple switches (A and B) in series.
Copy the table below. (It is called a truth table.)
Use the circuit to fill in the missing information.
Remember that **off = 0** and **on = 1**.

Switch A	Switch B	Bulb
0	0	0
0	1	?
1	0	?
?	?	1

2   Build this circuit. Make up a similar truth table for switches A and B and the bulb.

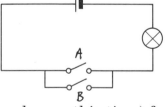

3   Try to make a truth table (four columns this time) for each of these two circuits **without** building them.

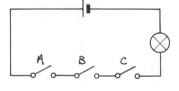
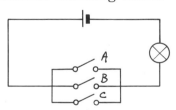

A radio receiver

Radio stations broadcast on different frequencies. You may know your local station's frequency. The frequency of transmission is the number of radio waves transmitted every second. It is measured in millions of waves per second (megahertz, MHz).

A radio receiver detects radio waves and changes them into sounds that you can hear. Three important parts of every radio receiver are

- the **tuner**
- the **amplifier**
- the **loudspeaker**

Low-frequency
transmission
(e.g. medium wave)

High-frequency
transmission
(e.g. VHF)

The tuner contains

- a **tuning capacitor:** this is what you turn when you tune in the radio to your favourite station
- an **aerial:** this is what collects the radio waves and changes them into an electrical current. A metal telescopic aerial outside the radio case is used for short wave, VHF (or FM) stations. A ferrite-rod aerial inside the case is used for long-wave and medium-wave stations.

The amplifier contains an electronic circuit which changes the small current from the aerial into a larger current. It will probably include transistors or integrated circuits.

The loudspeaker changes the electrical current into sounds that you can hear.

Collect

Labelled components
Radio
Aluminium foil

1 Examine the tuning capacitor, the ferrite-rod aerial and the loudspeaker on display.
   Find these parts inside the open radio.
2 Draw a rectangular box to represent the open radio. Label the positions of its main parts.
3 Use the radio to tune in to Radio 1 on medium wave.
   a With the volume quite low, turn the radio, and so the ferrite-rod aerial, around. Notice whether there is any difference in the sound.
   b Wrap the radio in aluminium foil while it is still on. Notice whether there is any difference in the sound.

continued ▶

continued ▶

1 Describe what each of the following parts of a radio receiver does:
   a the tuning capacitor
   b the ferrite rod
   c the amplifier
   d the loudspeaker.
2 What conclusion can you make about the ferrite-rod aerial from experiment 3a?
3 Are radio waves able to travel through aluminium foil? Explain your answer.

Collect

Flat pieces of solid

Design an experiment
   Find out if selected solids allow radio waves to pass through them.
1 Use a **flat** piece of each solid. Do not break or bend the piece of solid. The radio should be tuned to a medium wave station and left on at a **low** volume.

Hints
   ● Discuss the experiment with a partner.
   ● Find a way to use the flat piece of solid.
   ● Make sure you design a **fair** experiment. Each solid should be compared fairly with the others.
2 Design and carry out your experiment.

Write a report about the experiment that you designed. It should mention
   ● your method (draw a diagram if possible)
   ● your results
   ● your conclusions about which solids allow radio waves to pass through.

Biotechnology with yeast

The process of brewing beer uses yeast microbes to ferment sugar into alcohol. This is traditionally done in a fermenting vessel with one batch of product being made at a time. The vessel must be cleaned before the next fermentation.

In some processes it is now possible to fix the microbes in one place and use them over and over again. For example, yeast can also be used to change raw cane sugar (sucrose) into glucose. The glucose can then be used to make sweets. This process will run continuously with the same microbes being used again and again.

Collect

Tray of equipment

1 Clean all glassware and the bench with sterilising solution. Wash your hands.
2 Test the sucrose solution with Clinistix. (Use the index to find the method.) Check that there is no colour change.
3 Fix the yeast cells in small jelly beads by following these instructions.

a Stir 1 g yeast granules into 25 cm³ water.

b Add this to 25 cm³ sodium alginate solution. Stir all the time.

c Pour 200 cm³ calcium chloride solution into a third beaker

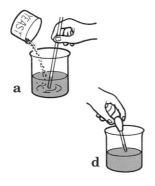

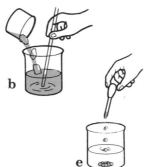

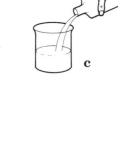

d Suck up some yeast/alginate mixture with a pipette.

e Drip this mixture from about 10 cm into the beaker of calcium chloride solution. Make lots of beads. *continued* ▶

▶ *continued*

4 Now set up your production line.

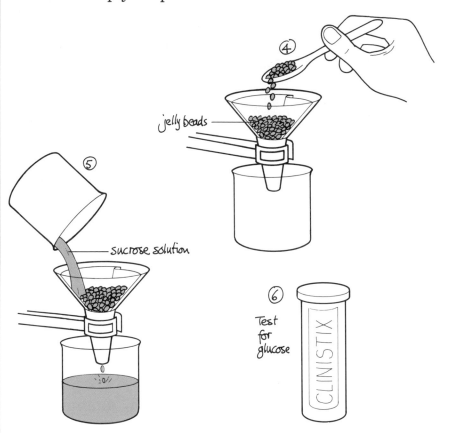

jelly beads

sucrose solution

Test for glucose

CLINISTIX

5 Pour the sucrose solution through the filter funnel twice.
6 Test a sample of the liquid you collect for glucose. If there is no change, pour the solution through the filter funnel once again and test it again.
7 If you have time, design a way of speeding up this reaction.

1 Explain the difference between a *batch* and a *continuous* process.
2 Imagine that you own a sweet factory! Why would yeast microbes be useful to you if sucrose was much cheaper than glucose?
3 Use the diagrams above to describe how your experimental production line worked.
4 What changes would you have to make to the experimental line if you were going to use it in a factory?

Controlling microbes at home

Home cleaning liquids often contain chemicals that are **disinfectants**. These kill microbes or prevent their growth. You are going to compare how well four household cleaners control yeast growth. The yeast is already growing on agar in a petri dish. The agar supplies the yeast with food.

Collect

Yeast agar plate
Cork borer
Tweezers
Cleaning liquids

1 Clean the bench and wash your hands.

2 Remove four discs of agar from the petri dish.

3 Carefully fill each well with a different cleaning liquid.

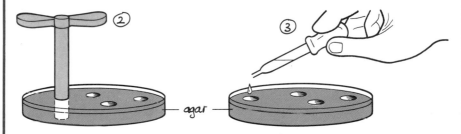

The chemicals will spread through the agar. There will be a clear patch around the chemical wells if the yeast cannot grow. The bigger the clear patch the better the chemical works.

4 Leave the petri dish for a day or so in a warm room.

Write a report about your investigation.
Use a coloured diagram to show where the yeast is growing well.

Waste not . . .

Recyclable

Materials are used in many different ways. Wood is used to make houses, fences and so on. Metal is used to make cars and tools. Plastic is used to make packaging and objects. After the material has been used it may be recycled, left to decay (rot), or destroyed.

Some materials can be recovered and used again. Metals, for example, are expensive to make. So scrap metal is often **recycled**. Glass and paper are sometimes recycled.

Some materials can be thrown away without harming the environment. Wooden objects like a clothes peg will decay. They are **biodegradable** because creatures like bacteria can eat them. Decay only occurs when the temperature, moisture and packing of the material are right for the microbes.

Some materials have to be **buried** or **destroyed**. For example, most plastics are durable – they will not usually rot because bacteria cannot eat them. Plastic waste stays around for a long time, so it is buried or burned in special furnaces.

Biodegradable

Durable

Collect
Sample rubbish bag

1 Carefully empty the contents onto the bench.
2 Divide the contents into three separate groups:
   ● things that can be recycled
   ● things that are biodegradable
   ● things that must be buried or destroyed.

1 Write down the meaning of *recycle* and *biodegradable*.
2 What conditions are needed for decay to take place?
3 Make a table to show the results of the rubbish bag survey.
4 Draw a bar graph to show how many items in the rubbish bag were in each of the three groups.

Picture quiz

1 **Collect** a copy of the cartoons drawn above. Stick them into your book.
   Give each cartoon a title.
2 Fill in the box below each cartoon by describing how microbes are involved.

Index

Acknowledgements

CARTOONS: Ainslie MacLeod; LINE DRAWINGS: RDL Artset
NATURAL HISTORY ARTWORK: Nancy Sutcliffe
COVER PHOTO: Kay Chernush/
The Image Bank

The following have provided photographs or given permission for copyright photographs or articles to be reproduced:

p.9 *top left* ZEFA; *top right* W. Broadhurst/Frank Lane Picture Agency; *centre left* John Lythgoe/Planet Earth Pictures; *centre right* Ivor Edmonds/Planet Earth Pictures; *bottom left* ZEFA; *bottom right* Geoff du Feu/Planet Earth Pictures

p.11 *far left* Patrick Clement/Bruce Coleman Ltd; *centre left* Eric Crichton/Bruce Coleman Ltd; *centre right* ZEFA; *far right* Mark Boulton/Bruce Coleman Ltd

p.13 *left* Mr and Mrs R. P. Lawrence/Frank Lane Picture Agency; *centre* A. Krumins/NHPA; *right* Andrew Henley/Biofotos

p.14 *top left* ZEFA; *top centre* F. W. Lane/Frank Lane Picture Agency; *top right* Joe Van Wormer/Bruce Coleman Ltd; *bottom left* David Phillips/Seaphot Ltd: Planet Earth Pictures; *bottom right* M. J. Thomas/Frank Lane Picture Agency

p.18 *a* Terry Whittaker/Frank Lane Picture Agency; *b* David Woodfall/NHPA; *c* GSF Picture Library; *d* ZEFA; *e* David Woodfall/NHPA; *f* J. Mackinnon/Bruce Coleman Ltd

p.19 *top left* N. G. Blake/Bruce Coleman Ltd; *bottom left* A. J. Deane/Bruce Coleman Ltd; *top right* Charles Henneghien/Bruce Coleman Ltd; *bottom right* Adrian P. Davies/Bruce Coleman Ltd

p.22 *top* Chris Newton/Frank Lane Picture Agency; *bottom left* Martin Bond/Science Photo Library; *bottom centre* ZEFA; *bottom right* ZEFA

p.23 *top* ZEFA; *bottom* GSF Picture Library

p.24 Walter Murray/NHPA

p.27 *left* ZEFA; *centre* GSF Picture Library; *right* US Department of Commerce/Frank Lane Picture Agency

p.28 GSF Picture Library

p.30 *magnesium* Russ Lappa/Science Photo Library; *sodium* GSF Picture Library; *potassium* GSF Picture Library; *titanium* Russ Lappa/Science Photo Library; *iron* GSF Picture Library; *chromium* Russ Lappa/Science Photo Library

p.31 *carbon* GSF Picture Library; *silicon* GSF Picture Library; *sulphur* GSF Picture Library; *bromine* Andrew McClenaghan/Science Photo Library; *iodine* GSF Picture Library

p.38 *left* ZEFA; *right* Warren Williams/Planet Earth Pictures

p.41 *left* ZEFA; *top right* H. Merton/Bruce Coleman Ltd; *bottom right* ZEFA

p.52 St Bartholomew's Hospital/Science Photo Library

p.61 Health Education Authority

p.62 *left* A. B. Dowsett/Science Photo Library; *centre* Dr Jeremy Burgess/Science Photo Library; *right* CNRI/Science Photo Library

p.74 *left* GSF Picture Library; *right* TRRL

p.76 Tony Baker/Haymarket

p.78 *top* CNRI/Science Photo Library; *bottom* James Stevenson/Science Photo Library

p.91 Ronald Rover/Science Photo Library

p.99 *top* Norman Tomalin/Bruce Coleman Ltd; *bottom* GSF Picture Library

p.100 *top* Dr Jeremy Burgess/Science Photo Library; *centre* Mansell Collection; *bottom* NASA/Bruce Coleman Ltd

p.104 *top left* ZEFA; *centre left* Dick Luria/Science Photo Library; *bottom left* AGA Infrared Systems/Science Photo Library; *right* Hank Morgan/Science Photo Library

p.105 Alfred Pasieka/Science Photo Library

p.107 GSF Picture Library

p.108 *left* ZEFA; *centre* Alfred Pasieka; *right* Frans Lanting/Bruce Coleman Ltd

p.109 BBC

p.114 *top* St Mary's Hospital Medical School/Science Photo Library; *bottom* John Durham/Science Photo Library

p.122 Prof. David Hall/Science Photo Library

p.124 *top left* David Parker, 600 Group Fanuc/Science Photo Library; *bottom left* Dale Boyer/NASA/Science Photo Library; *top centre right* Alexander Tsiaras/Science Photo Library; *top far right* Roger H. Allen/Barnaby's Picture Library; *centre right* Dr Ian Robson/Science Photo Library; *bottom right* Colorsport

p.125 Aileen Ballantyne/*The Guardian*

p.126 Institute of Child Health; *New Scientist*; Harrow Health Authority; Helix; Scottish Wildlife Trust

p.128 Gunter Ziesler/Bruce Coleman Ltd

p.133 David Woodfall/NHPA

p.135 *left* David Parker/Science Photo Library; *right* E. A. Jones/NHPA

p.137 *left* ZEFA; *right* ZEFA

p.139 *left* Eric Crichton/Bruce Coleman Ltd; *right* GSF Picture Library

p.149 Bill Sanderson/Science Photo Library

p.151 GSF Picture Library

p.155 ZEFA

p.157 *all photos* GSF Picture Library

p.159 *top far left* Stammers/Science Photo Library; *top second left* Griffin & George; *top right* GSF Picture Library; *bottom* Adam Hart-Davis/Science Photo Library

p.172 *upper left* Alex Bartel/Science Photo Library; *lower left* Brian Hawkes; *lower right* Dr Morley Read/Science Photo Library